THIRTEEN

THIRTEEN
The Apollo Flight That Failed

Henry S. F. Cooper, Jr.

THE JOHNS HOPKINS UNIVERSITY PRESS
Baltimore and London

Printed in the United States of America on acid-free paper

The material in this book appeared originally in *The New Yorker* magazine in slightly different form.
Published in a hardcover edition, *Thirteen: The Flight that Failed,* by Dial Press, 1972
Johns Hopkins Paperbacks edition, 1995
04 03 02 01 00 99 98 97 96 95 5 4 3 2 1

The Johns Hopkins University Press
2715 North Charles Street
Baltimore, Maryland 21218-4319
The Johns Hopkins Press Ltd., London

Library of Congress Cataloging-in-Publication Data will be found at the end of this book.

A catalog record for this book is available from the British Library.

ISBN 0-8018-5097-5 (pbk.)

TO THE EDITORS AND STAFF OF
THE NEW YORKER

ACKNOWLEDGMENTS

I WANT TO THANK THOSE ASTRONAUTS AND FLIGHT controllers of Apollo 13 who gave me their stories and who later took the time to make sure I got them straight.

Many others have had a hand in this book. In particular, William Shawn, the Editor of *The New Yorker,* first had the idea that the Apollo 13 mission might offer the best glimpse into the anatomy of a spaceflight and especially into the workings of "those men who sit at those desks" in the Mission Control Room—the flight controllers, whom nobody seemed to know much about. Later, William Knapp and Anne Britchky wrestled with proofs and made sure everything was in order.

Several others helped me at the Manned Spaceflight Center outside Houston, among them Milton Reim and Terry White of the Public Affairs

Office, who aimed me in the right direction and made sure the doors were open when I got there.

Finally, my thanks go to Robert Cornfield at The Dial Press, who must have wondered from time to time whether there would be a manuscript at *all.*

New York, N.Y. Henry S. F. Cooper, Jr.
December 1972

OUT

AT A LITTLE AFTER NINE CENTRAL STANDARD TIME ON the night of Monday, April 13, 1970, there was, high in the western sky, a tiny flare of light that in some respects resembled a star exploding far away in our galaxy. At the Manned Spacecraft Center, near Houston, Texas, the glow was seen by several engineers who were using a rooftop observatory to track the Apollo 13 spacecraft, which had been launched two days before and was now a day away from the moon and two days from a scheduled moon landing. One of the group, Andy Saulietis, had rigged a telescope to a television set in such a way that objects in the telescope's field of view appeared on the screen. Above, the sky was clear and black, like deep water, with occasional clouds making ripples across it. Saulietis and his companions —who, incidentally, had no operational connection with the Apollo 13 mission but were following it for a related project—had lost sight of the spacecraft, two hundred and five thousand miles away. However, they had been watching the much larger booster rocket that had propelled the spacecraft out of earth orbit and was trailing it to the moon; the booster had appeared as a pinprick of light that pulsed slowly, like a variable star, for it tumbled end over end as the result of dumping its fuel, and the sunlight glinted off it with varying intensity. Shortly before nine, the observers on the rooftop at Houston had lost track of the booster, too, for the pinprick had been almost at the limit of the resolving power of their equipment. Suddenly, near the middle of the TV screen, a bright spot appeared, and over the next ten minutes it grew to be a white disc the size of a dime. The rooftop watchers had no communications link with the Mission Control Center, about two hundred yards away—a large building consisting of two linked wings, with operations rooms in one, offices in the other—and they had no reason to connect the flaring light with the spacecraft or to be concerned with the safety of its crew: Captain James A. Lovell, Jr., of the United States Navy, who was in command; and the pilots for the command and lunar modules, John L. Swigert, Jr., and Fred W. Haise, Jr., both civilians. It was to be some time before either Mission Control or the astronauts themselves realized that one of the ship's two oxygen tanks had burst, spewing into space three hundred pounds of liquid oxygen, which meant the loss of half the craft's supply of this element for generating electricity and water. The oxygen came out in one big blob, and in gravityless space it formed a gaseous sphere that expanded rapidly; the sunlight made it glow. In ten minutes, it was thirty miles in diameter. Then the white disc slowly disappeared—though traces of it were observed an hour later through a more powerful telescope in Canada.

Saulietis and the others assumed the white spot to be a defect in their television set, which had been flickering and blipping badly, so they went home to bed and thought nothing about the incident until the next morning. They were not the only ones who failed to grasp the situation. After two successful lunar landings, which had been preceded by two Apollo flights around the moon, no one at the Space Center was thinking in terms of accidents. Later, some of the first interpretations of what had happened would center on the notion that the spacecraft had been struck by a meteor—a borrowing from science fiction, for Jules Verne's space capsule in *From the Earth to the Moon* had almost been hit by one while it was approaching the moon. No one believed that there could be any flaw in the craft itself. Yet in the Mission Control Center, where dozens of automatic pens were scribbling data radioed from Apollo 13, at the time of the explosion the pens stopped writing for almost two seconds—a "drop-out of data" indicating a major problem with either the electrical system or the system of transmitting data from the craft. No one noticed.

Inside the Mission Control Center's Operations Wing, a chunky, monolithic three-story structure as white and silent as a block of ice—the geometric representation of an intelligence brooding on far-off space—the flight controllers were at least as thoroughly cut off from the world around them as sailors belowdecks on a ship at sea. There were no windows, as there are none on the lower decks of a ship—and, in a sense, the Operations Wing really *was* the lower decks of a ship, the upper deck being a couple of hundred thousand miles away in space. Astronauts are more like officers aboard a large ship than like solitary heroes, and that may account for some of the difficulty many people have in comprehending their roles. In the spacecraft and the Mission Control Center combined, there were about as many astronauts and flight controllers as there are officers aboard a big vessel, and they worked together as closely as officers belowdecks work with officers on the bridge. In fact, one of the controllers, the Flight Director, in some respects might have been regarded as the real skipper of the spacecraft, for although the relationship between the astronauts and the ground crew was a delicate, interdependent one, the astronauts usually did what he advised, particularly in an emergency. Though the flight controllers were on earth, they had, by means of telemetry—data radioed constantly from the spacecraft—more information about what was happening aboard than the astronauts themselves had. The walls of the Mission Operations Control Room, on the third floor of the

Operations Wing, were the same color as the inside of the command module—gray. Five big screens at the front of the room might have been windows looking out into space: the middle screen showed the earth on the left and the moon on the right, with a bright-yellow line representing the spacecraft's trajectory as it lengthened slowly between them. And some of the consoles at which the flight controllers sat duplicated equipment aboard the spacecraft.

The flight controllers, most of whom were in their twenties and thirties, sat in four rows. Before the accident, they were relaxed—even bored. The first fifty-five hours of the flight had gone so smoothly that they had once sent word to the astronauts that they were "putting us to sleep down here." One team of flight controllers had been reduced to commenting on the number of "thirteens" that cropped up; for example, the time of launch in Houston—the official time for the flight—had been 13:13, or in our terms 1:13 P.M. About the only event requiring the controllers' close attention since Apollo 13 left earth orbit had been a small rocket burn the day before, called a "hybrid transfer maneuver," which had aimed the spacecraft for its target on the moon, the Fra Mauro hills—and, incidentally, had taken it off a free-return trajectory, the safe path most previous Apollos had followed so that in the event of trouble the spacecraft could, without navigational adjustment, swing around the moon and head back to earth. Now, a couple of minutes before nine, one of the flight controllers, the Retrofire Officer, whose responsibility it was always to have a plan ready, in case of trouble, for bringing the astronauts home, sent word to the Flight Director that another bridge was about to be burned: via a pneumatic-tube system connecting the consoles, he dispatched a routine memorandum to the effect that the spacecraft was nearing the spot where it could no longer reverse its direction and return directly to earth if anything went wrong.

There were a couple of dozen controllers on duty, of whom only about half were directly involved with the running of the spacecraft at any given time. In the front row, which was called the Trench, sat three Flight Dynamics Engineers, the men responsible for the ship's trajectory: from right to left, the Guidance Officer, or GUIDO, who was the chief navigation officer; the Flight Dynamics Officer, or FIDO, who plotted the trajectory and made sure the spacecraft followed it; and the Retrofire Officer, or RETRO, who was in charge of the spacecraft's reëntry into the earth's atmosphere. Most of the second row, behind the Trench, was taken up by the Systems Operations Engineers, who monitored the equipment inside the spacecraft: in the center, the EECOM, who looked after the electrical, environmental, and other systems in the command module, where the astronauts rode; next to him, the LM Systems Officer, or TELMU, who did the same thing for the lunar module, in which the astronauts would land on the moon; and then two Guidance and Navigation Control Officers—one, the GNC, for the command module, and the other, the CONTROL, for the lunar module—who were in charge not only of the guidance and navigation equipment in the two modules but of the propulsion systems as well. To their left was the Spacecraft Communicator, known—from the old Mercury-capsule days—as the CAPCOM, the only man who could talk directly with the astronauts, and to his left was the Flight Surgeon. Behind the Flight Surgeon, in the third row, was the Instrument and Communications Officer, or INCO, who was responsible for the radios and telemetry transmitters aboard the spacecraft. Finally, in the center of the third row— a good vantage point for keeping an eye on everyone else—there was the Flight Director, the ship's earthbound co-captain. (In the fourth row sat administrators, including a Public Affairs Officer.)

There were four shifts, or teams, of flight controllers—White, Black, Maroon, and Gold—and at that moment the White Team was on duty. The controllers talked to each other over an intercom hookup called the loop. To cut down on what they called loop chatter, which had a way of sounding like random thoughts popping up in a single individual's mind, the controllers referred to each other by their acronyms or abbreviations: FIDO, GUIDO, RETRO, CAPCOM, EECOM, and so on. Apollo 13 was just plain "Thirteen."

On April 13, about half an hour before the white spot was seen by Saulietis and his companions, the flight controllers were watching a television show, which the astronauts were broadcasting from the spacecraft, and which was projected on one of the big screens at the front of the room. As it happened, none of the three major networks carried the telecast—though they would show tapes of it later—and it concluded ten minutes before the occurrence of the episode that could have made it as dramatic as any performance in history. As the flight controllers leaned back in their chairs to watch, they thought the astronauts seemed happy. Captain Lovell, the commander, who was forty-two, and who had graduated from Annapolis in 1952, ten years before he became an astronaut, was cameraman and announcer for the show; he first panned the camera around the gray interior of the command module, a cone whose base was almost thirteen feet in diameter and whose height was ten and a half feet. It was about as big as the inside of a small station wagon, though the astronauts, who could float about, found it roomier than a similar space on the ground. Lovell was resting on the left couch. Beneath him was the service module, which contained, among other things, the electrical system, including the two big oxygen tanks. The cylindrical service module, with the conical command module at one end, formed a single pointed unit, in front of which was the bug-like lunar module, giving the spacecraft a total length of almost sixty feet. Above Lovell's head, at the apex of the cone, was a round hatch leading through a short tunnel to the lunar module. Lovell had flown in space three times, and this was the second time he had set out for the moon; he had circled it, in December, 1968, as a member of the Apollo 8 mission. This may have accounted for a certain bland professionalism he displayed as master of ceremonies. He began, "What we plan to do for you today is start out in the spaceship Odyssey and take you on through from Odyssey in through the tunnel into Aquarius." Odyssey was the code name for the command module, and Aquarius for the LM. (The latter was named for a song in the musical "Hair," which Lovell—who was Special Consultant to the President's Council on Physical Fitness and Sports—had not seen. When he caught up with the show later, he walked out.)

Lovell aimed the camera at Haise, the lunar-module pilot, who was hovering by the hatch, ready to lead the way into the LM. Clothed, like the others, in white coveralls, Haise was hard to make out, because the television relay was none too sharp. Haise, a native of Biloxi, Mississippi, is a slight man with dark-brown hair and a square jaw, who speaks with a slight drawl. Although he had become an astronaut only four years before, and this was his first spaceflight, he had made enough of an impression so that Lovell and the other Apollo 8 astronauts had seen fit to name a crater on the moon after him. Haise was not particularly busy at the moment—the LM was not scheduled to be powered up until they were in lunar orbit, a day hence—so Lovell had persuaded him to act as guide for his tour, and now Lovell, holding the TV camera at arm's length, followed Haise as he swam through the tunnel into the LM. There Haise demonstrated various pieces of equipment to be used on the moon, including a rectangular bag (called the Gunga Din) that he and Lovell would wear inside their helmets, so that they could drink while they walked about the Fra Mauro hills. "So if you hear any funny noises on television during our moon walk, it is probably just the drink bag," Lovell said. Haise was doing something in the middle of the LM now, but the flight controllers had trouble seeing exactly what it was. The CAPCOM asked if he were opening the food locker, and the flight controllers laughed, because Haise had a well-known penchant for food. Haise said that he was rigging his hammock for sleep, and the CAPCOM replied, "Roger. Sleeping and then eating."

Leaving Haise in the LM, Lovell went back through the tunnel and, in the command module, sought out the third member of the crew—Swigert, the command-module pilot. "There he is! We see him!" the CAPCOM said, and, sure enough, there he was, seated before the ship's controls in the middle one of the astronauts' three seats, surrounded on three sides by nine dashboard panels. Swigert, a sharp-faced, sharp-eyed man, who was born in Denver, Colorado, in 1931, and had become an astronaut with Haise in 1966, was too busy just then to do more than smile at the camera. He did not take a big part in the television show. This was his first spaceflight, as it was Haise's, but he felt himself to be under more pressure, because he had been merely the backup command-module pilot, and had been officially assigned to the flight only the day before it left, after the prime crew had been exposed to German measles and it was discovered that the prime command-module pilot was susceptible. Lovell had worked with Swigert for two days without letup before agreeing to take him on the flight. (Swigert had been in such a rush that he hadn't thought about completing his 1969 income-tax return, due in four days, until he was a quarter of the way to the moon.) During the first fifty-five hours of the flight, Swigert had run into a few minor difficulties; for instance, he had been having trouble reading the quantity gauge for one of the oxygen tanks, which had gone off-scale on the high side, and earlier that day the CAPCOM had told him he could expect frequent requests from the ground to turn on the fans in the tank to stir up the oxygen —what is called a "cryogenic stir"—for the purpose of obtaining accurate quantity readings. Now, as the television show continued, Swigert found a moment to hold the camera and trained it on the commander; Lovell appeared on the screen for the first time. A tall, sober-looking man whose face at times breaks into a broad grin, Lovell demonstrated a tape recorder that could play a number of songs, among them "Aquarius" and the Richard Strauss "Thus Spake Zarathustra" theme used in the film *2001: A Space Odyssey.* At length, the CAPCOM broke in to suggest that Lovell conclude the program. The commander replied, "Roger. Sounds good. This is the crew of Apollo 13 wishing everyone there a nice evening, and we're just about ready to close out our inspection of Aquarius and get back for a pleasant evening in Odyssey. Good night." It was then 9:00 P.M.

As the show ended, Lovell joined Swigert at the controls in the command module, sitting in the left-hand seat, and helped him copy down an instruction from the CAPCOM for rolling the spacecraft to the right in order to photograph a comet named Bennett. In front of them were two red lights labelled "Master Alarm," which would flash on if the spacecraft computer detected a serious malfunction, and over their heads was an array of yellow caution lights to indicate minor malfunctions. One of these flashed on at five minutes past nine, and so did a similar one in Houston on the console of the EECOM, who was in charge of monitoring, among other items, the spacecraft's equipment for generating electricity. The EECOM on duty was Seymour Liebergot, a thirty-four-year-old electrical engineer from California State College at Los Angeles. The yellow light warned of low pressure in a hydrogen tank in the service module, which was crammed with equipment; in addition to the main propulsion system, there was the system for generating water and electricity, of which the balky hydrogen tank was a part. The generating system was simple and efficient: hydrogen and oxygen reacted inside units called fuel cells to generate electricity and, at the same time, to produce most of the spacecraft's water. Liebergot wasn't worried by the alarm, because the system was redundant: there were two hydrogen tanks, two oxygen tanks, and three fuel cells, and if anything went wrong gases could be routed from any of the tanks to any of the cells.

Liebergot had been manipulating the hydrogen quantity in the tanks all along, so the hydrogen warning was almost routine. It did, however, preëmpt the circuits of the warning system, so that a problem with the oxygen supply would not turn on a yellow light, as it was supposed to. To make sure he was getting the right information, Liebergot asked the Flight Director, Eugene Kranz, to get the astronauts to stir the hydrogen in both tanks; even though Kranz was only four feet behind him, Liebergot had to get his attention over the loop. Kranz, who was thirty-six, and is a graduate of St. Louis University, had been with NASA since 1960. As the Flight Director on duty, he was in charge of everything that went on in the Control Room and therefore had to approve all requests of this sort before they could be relayed to the spacecraft. Also, as Chief of the Flight Control Division, he was in over-all charge of the controllers. Liebergot now asked that in addition to the hydrogen tanks the oxygen tanks be stirred, for, like Swigert up in the CM, he had been having trouble all day getting an accurate reading on the quantity of Oxygen Tank No. 2.

If Liebergot had been able to look inside the two oxygen tanks, he would not have wanted to risk disturbing them. They, together with the rest of the electrical-generating system, were inside Bay 4 of the service module—one of the six compartments that ran the length of the twenty-five-foot module. The interior of Bay 4, a place of silvery insulation and golden wires, was divided into compartments. The three fuel cells were in the forward one; the two hydrogen tanks were in the rear; and in the middle were the oxygen tanks, two silvery spheres, twenty-six inches in diameter and made of a tough nickel-steel alloy called Inconel, which were strong enough to contain the oxygen under nine hundred pounds of pressure per square inch. They had an outer and an inner shell, and the space between was filled with insulation—some of it inflammable. On top of each tank was a capped dome that sealed an opening for pipes and for wires that brought electricity to instruments inside the tank—fans, heaters, and the sensors for the quantity, temperature, and pressure gauges.

What would have alarmed Liebergot if he had been able to look inside the tanks was that the wires in Oxygen Tank No. 2 were largely bare of insulation. This situation, attributable to both imperfect design and human inattention, had existed for more than two weeks—since March, after a ground crew at Cape Kennedy had piped liquid oxygen into the tanks in a countdown demonstration test. The oxygen and the hydrogen were cryogenic, or cooled to a liquid state, in order to keep them at sufficiently low volume so that they could be compactly stored; the temperature of the oxygen at filling time was two hundred and ninety-seven degrees below zero Fahrenheit. When the test was over, the engineers had been unable to get the oxygen out of Tank No. 2. This trouble may have arisen because the cap on top of the tank had been jolted so that pipes and wires inside were loosened—which may also be why Liebergot was now having trouble with his tank readings. In any event, the ground crew had tried to force the oxygen out by turning on the heaters and fans inside the tank; the fans would stir up the oxygen and the heaters would warm it to make it expand, thus expelling it. The heaters were left on for eight hours—a longer period than such heaters had ever been on before—and during that time nobody was aware that the temperature inside the tank was getting higher and higher. The ground crew was not worried, because they knew there was a thermostatic safety switch in the tank's interior that was supposed to turn the heaters off if the temperature rose above eighty degrees, the safe limit. So confident were the designers of the equipment that although they provided a thermometer to give a temperature reading from inside the tank, this did not register above eighty-five degrees. What the designers had disregarded, though, was that the safety switch they specified was built to operate on the twenty-eight-volt current of the spacecraft's power supply, and that when the tanks were tested at Cape Kennedy they were powered from a sixty-five-volt supply. The designers had thought the switch would be kept cool during tests by being immersed in the supercooled oxygen. Instead, as Tank No. 2 emptied, the safety switch overheated and failed. As was determined much later by experimentation with similar equipment under similar circumstances, the switch undoubtedly fused shut so that it couldn't turn off the heaters. The failure could have been discovered had any of the ground crew noticed that the heaters were still drawing current for hours after they should have turned off, and thus were still in operation; apparently, no one looked at the current gauge. The heat might well have gone up to a thousand degrees—enough to burn the insulation off the wires. After that, if electrical equipment inside the tanks was turned on and the wires happened to come close together, a spark could pass between them.

When Liebergot requested the cryogenic stir, Kranz, the Flight Director, said he would like to hold off awhile on relaying the message, because he wanted to give the astronauts time to settle down after their TV show. Kranz, a big-boned, trim man with fair hair cropped so close that from certain angles it was barely visible, often suggested a tough Marine Corps unit commander. He was wearing a flashy, iridescent white vest, in honor of his team of flight controllers, the White Team. In about an hour, Kranz's team would be handing over control to the Black Team, and already Black Team controllers had begun to arrive and draw up chairs so that they could look over the shoulders of their White counterparts.

Kranz was delaying action on Liebergot's request until Haise had returned to the command module. To determine whether he had started back yet, Kranz asked the TELMU, who monitored the data on the LM, whether the LM's hatch was still open. The TELMU thereupon checked the electrical-power readings for the command module, from which the inert LM was drawing the small amount of power it needed. By pressing a combination of buttons in front of him, each of the flight controllers could throw onto a small television screen in front of him any one of two hundred and fifty charts giving data on the spacecraft's condition; these were prepared by computers on the ground floor of the Operations Wing, where the telemetry was received. There had been a slight drop in the power output, so the TELMU guessed that the LM's hatch was closed—since the lights that turned on when it opened were now drawing no electricity—and therefore that Haise was on his way back to the command module. Accordingly, Kranz told the CAPCOM to radio to Swigert, the command-module pilot, the message to stir the tanks. Swigert, looking up into the apex of the command module, could see Haise coming back through the short tunnel, and that was about all he could see, for he was hemmed in by over five hundred dials, buttons, knobs, switches, and thumbwheels. Most of them were guarded by little U-shaped wickets, lest an astronaut bump against one inadvertently. Swigert's movements were gingerly; as the new crew member, he was especially anxious to perform as he should. When he received Liebergot's message, he pressed four switches to his right. In the Control Room, Liebergot sat forward to get a better look at the screen on his console, which would now show the pressure, quantity, and temperature readings of the tanks.

Nothing much happened for sixteen seconds. Then, inside Oxygen Tank No. 2, an arc of electricity shot between two naked wires. In the next twenty-four seconds, the arc heated the oxygen, and its pressure rose rapidly. Because the hydrogen-tank low-pressure signal had preëmpted the system, no caution lights flashed, and because Liebergot was concentrating on the readings for the hydrogen tanks, which were on the right side of his television screen, he didn't notice the rapidly increasing numbers in one of the oxygen-pressure columns, three inches to the left of where he was looking. During the time the pressure in the oxygen tank was increasing, the only person in the Control Room to notice that anything was wrong was William Fenner, the GUIDO, who saw what he called an "event"—an unexpected number—on his console. It signalled what he called a "hardware restart," which meant that the spacecraft computer had found a problem and was going back over recent events to find out where the trouble lay. It never found out, but the restart provided Kranz with a false trail to follow later.

On the basis of recorded data, of evidence brought back by the astronauts, and of extensive post-mission analyses, it is possible to reconstruct with a fair degree of certainty what happened during this two-minute period. At the end of twenty-four seconds—at eight minutes past nine—the oxygen pressure had blown the dome off the top of the tank. The layer of insulation between the inner and outer shells of the tanks undoubtedly caught fire, with flames, fanned by the rush of escaping oxygen, spewing as from a blowtorch all over the inside of Bay 4 of the service module. The silvery sheets of Mylar insulation—heat-resistant but nevertheless inflammable—lining the inside of the bay probably caught fire, and the resulting gases blew out the bay's cover, which was one of the six panels making up the service module's external hull. It was lucky the panel blew out when it did, for if the pressure had been allowed to build up much more, the command module itself, plugging the front end of the service module like a cork, could have blown off instead. Later, in describing what happened, NASA engineers avoided using the word "explosion;" they preferred the more delicate and less dramatic term "tank failure," and in a sense it *was* the more accurate expression, inasmuch as the tank did not explode in the way a bomb does but broke open under pressure.

Whether called an explosion or a tank failure, such an event is less noticeable in space than it would be on the ground, where air transmits sound and shock waves. Therefore, none of the astronauts were aware that one of the oxygen tanks had ruptured. Nevertheless, each of them was instantly made aware, in one way or another, that there had been an untoward event. First, Swigert reported over the radio that they seemed to have a problem. His voice was so calm that the CAPCOM, Jack R.

Lousma, could not tell which of the astronauts was speaking, and Lousma knew the astronauts well, because he was an astronaut himself. What had disturbed Swigert, as he later recalled it, was not so much the sound of a perceptible bang as the sensation of a sort of shudder that ran through the spacecraft. He could not make a precise distinction, he said, because the borderline between feeling a vibration and hearing it is sometimes imperceptible. What he felt may in fact have been not unlike the disconcerting shudder that first puzzled some of the passengers aboard the *Titanic* as the ship scraped against an iceberg. Swigert was strapped into his seat, and so was better able to feel the shudder than Lovell, the spacecraft commander. Lovell, who was floating just above his seat, said later he had not felt the shudder but *had* heard a distinct bang. Lovell's first thought was that the bang had been made by Haise opening a valve in the lunar module. At thirty-six, Haise still looked like the youthful, irrepressible sort of person who might make a loud noise without warning. However, Haise was at this moment emerging from the tunnel, and Lovell could tell by the look on his face that he, too, had been jolted by something. Far from causing the bang, he had been startled when the tunnel shook up and down—a motion he thought ominous, for normally when the tunnel shook it was from side to side. He immediately felt that something fundamental was wrong.

Both Lovell and Swigert thought that the bang —or shudder—had come from the lunar module, and as Haise emerged from the tunnel Swigert shot out of his seat and slammed the command-module hatch shut behind him. Haise scrambled to his seat —the right-hand one—for the master alarm was now sounding in his earphones. Swigert had noticed an amber caution light glowing overhead. It didn't signal trouble in the oxygen tank, because that alarm system was still tied up by the low-pressure warning in the hydrogen tanks; rather, it signified trouble with the electrical system, the controls for which were near Haise. About this time, the Flight Surgeon, Dr. Willard R. Hawkins, noticed that the pulse readings for all three astronauts had shot up from about seventy to over a hundred and thirty.

The first disaster in space had occurred, and no one knew what had happened. On the ground, the flight controllers were not even sure that anything *had.* One reason for their ignorance was the imperfect nature of the telemetry from the spacecraft, which could not tell them directly that an oxygen tank had blown up. It could only report what the temperatures and pressures were in the tanks, whether certain voltages were within the proper limits, and whether certain equipment was on or off. This information had to be interpreted before the flight controllers could know what was going on, and the flight controllers were slow to make the correct interpretation, because, like everyone else at NASA, they felt secure in the knowledge that the spacecraft was as safe a machine for flying to the moon as it was possible to devise. Obviously, men would not be sent into space in anything less, and inasmuch as men *were* being sent into space, the pressure around NASA to have confidence in the spacecraft was enormous. Everyone placed particular faith in the spacecraft's redundancy: there were two or more of almost everything. Even the flight controllers' own training contributed to their confidence. For three months before the flight, they had flown the mission over and over in rehearsals called simulations. For these, a team of flight controllers took their places at the consoles in the Control Room while the astronauts got inside simulators—working models of the spacecraft, very much like the Link Flight Simulators that student airplane pilots use. Both groups were connected to computers that had been programmed to create problems likely to come up on the mission. The previous moon flights had gone so well that the flight controllers had complained on an earlier occasion to the men planning the simulations that these were too tough to be authentic. Accordingly, the simulations in preparation for Apollo 13 had dealt only with problems that were considered likely to arise; the controllers hadn't wasted time on what one engineer called "four-point failures—way-out disasters."

Before the flight controllers could admit the full scope of the present disaster, they went to great lengths to find explanations that would not involve a major failure of the spacecraft. It took them a quarter of an hour to get a rough idea of what had happened, and about an hour more to admit that the spacecraft was damaged beyond repair. At the outset, Liebergot, the EECOM, wasn't particularly alarmed. Because he had happened to miss seeing the sudden rise in pressure in Oxygen Tank No. 2, it simply didn't occur to him that the tank had blown out. There was such a cascade of problems that, not having noticed where they started, he didn't know where to begin to look for their source. Since he had no reason to think in terms of the oxygen tank in the first place, he had to track the trouble backward step by step all the way through the electrical system. The only clue he had to start with was the electrical warning Haise had reported and a similar light flashing on his own console. When he tracked it down, he found that it signalled what he called "a Main Bus B undervolt." A main bus is like a set of wall plugs. (Electricians also call it a distribution terminal board.) Electricity from the fuel cells—the generators—was fed into the buses, and then power was tapped out of them by the equipment that needed it. For redundancy, there were two main buses, A and B, and what Liebergot had found was that Main Bus B had suffered a significant drop in power, so that the equipment connected to it, which was half the equipment in the spacecraft, was in danger of failing. Up in the command module, Haise already had a sinking feeling, for, according to the mission rules, both buses had to be operating if the astronauts were to get the go-ahead to land on the moon.

Then there was a moment of relief. Haise saw the warning light above his head flicker out, and down in the Control Room the same thing happened on Liebergot's console. Lovell reported to the CAPCOM that the power in the bus was back to normal. Over the loop, Liebergot suggested to Kranz that the trouble might not have been an undervolt at all but, rather, a problem with the instruments reporting the problem. In the next hour or so, they came back over and over again to this wishful explanation—what flight controllers call an "instrumentation failure." Following this false trail, they told each other that perhaps everything was all right after all—though Haise now told the CAPCOM that "a pretty large bang" had been associated with the incident. Oddly, Kranz had not heard the astronauts mention a "bang" before. Now a light flickered on one of the panels on his console to indicate that one of the flight controllers—the INCO, who was in charge of the radios aboard the spacecraft—wanted to talk to him.

The INCO told Kranz about a communications "funny"—an aberration that doesn't clear up immediately, as opposed to a "glitch," which is a transitory one. At the time of the bang, the INCO reported, there had been an unexplained change in the width of the radio waves transmitted from the spacecraft: they had suddenly switched from a narrow beam to a wide one. Kranz was still not alarmed. The spacecraft radio was transmitting with the high-gain antenna—a sort of stick with reflectors that had to be aimed as precisely as a rifle—and it crossed Kranz's mind that since the antenna ran on power from Main Bus B, the undervolt might somehow have caused the change; if that was so, then the funny should correct itself now that the undervolt had. Much later, it became apparent that when the side panel of the service module had been torn off and hurtled into space it hit the antenna, causing a change in the nature of the radio signal.

Less than a minute had passed since the accident. A voice from the spacecraft now said that the bang must have affected the gauge that reported the level of Oxygen Tank No. 2—first it had oscillated between twenty and sixty per cent, but now it was off-scale on the high side. This still did not cause Kranz or Liebergot to think that there might be a problem with the oxygen tank. They had been having trouble with the oxygen gauges all along, and they thought that the same trouble had cropped up once more. It was hard for anyone to get rid of the idea that the instruments were lying to them. Just then, Lovell reported that Main Bus B had no power in it at all and Main Bus A was beginning to show an undervolt, too; that is, one main bus had gone dead and the other was losing power. If both buses died, the command and service modules would be without any electrical power except a small amount available from three storage batteries to be used during the return through the earth's atmosphere. Liebergot was confused. The two main buses, themselves paired for redundancy, were drawing their power from *three* redundant fuel cells; if one bus died, there was every reason for the other to hold up. While Liebergot pondered, there was a long silence, broken at last by Lovell, who asked, a little anxiously, "O.K., Houston, are you still reading Apollo 13?"

Lousma replied, "That's affirmative; we're reading you. We're still trying to come up with some good ideas here for you." Then, pressing a switch so that the astronauts couldn't hear him, Lousma said hurriedly to Kranz, "Is there any kind of lead we can give them, or are we looking at instrumentation problems, or have we got real problems, or what?"

About six minutes had passed since the accident.

Over the loop, Kranz asked Liebergot for recommendations. Kranz had once said that the hardest part of being a flight controller was being "the last man in the decision chain." That occurred, Kranz said, when a problem was passed to a flight controller at the last minute and he had to solve it all by himself; if he made a mistake, he did it in front of the whole world, and possibly jeopardized the mission. This was Liebergot's situation now. On his telemetry screen, he could produce far more information about the spacecraft's electrical system than the astronauts themselves had; the trouble was that he had no idea where to start looking. Half of the lights before him were amber, and he recalled that the only other time this had happened was shortly after lift-off during the Apollo 12 mission. He also recalled that on that occasion the spacecraft had been hit by lightning—a recollection that did him no good whatever, since there was no lightning two hundred and five thousand miles out in space. He now set about trying to piece together what he knew. The main question he had to answer was why there had been an undervolt in Main Bus B. Spread out before him on the console was a diagram of the spacecraft's electrical system. It made a sort of chain, leading from the hydrogen and oxygen tanks to the fuel cells and from those to the buses, and ending with the equipment in the spacecraft that was receiving—or was supposed to be receiving—power. At the moment, they were connected as shown in the following diagram (the hydrogen tanks are not included).

Fuel Cells 1 and 2 were supplying electricity to Main Bus A, which was still working. Main Bus B, however, was drawing its electricity from Fuel Cell 3 alone, and it didn't take Liebergot long to find out that Fuel Cell 3 had stopped generating power. Then he learned that Fuel Cell 1 was generating no power, either. It looked as if the spacecraft had lost two fuel cells—an unprecedented situation. The spacecraft was getting along solely on the electricity that Fuel Cell 2 was supplying to Main Bus A—and, inexplicably, the power put out by the one remaining good fuel cell was beginning to drop, too, though not yet enough for Liebergot to worry about.

The two apparently dead cells, Liebergot knew, drew their oxygen from the same two tanks, but

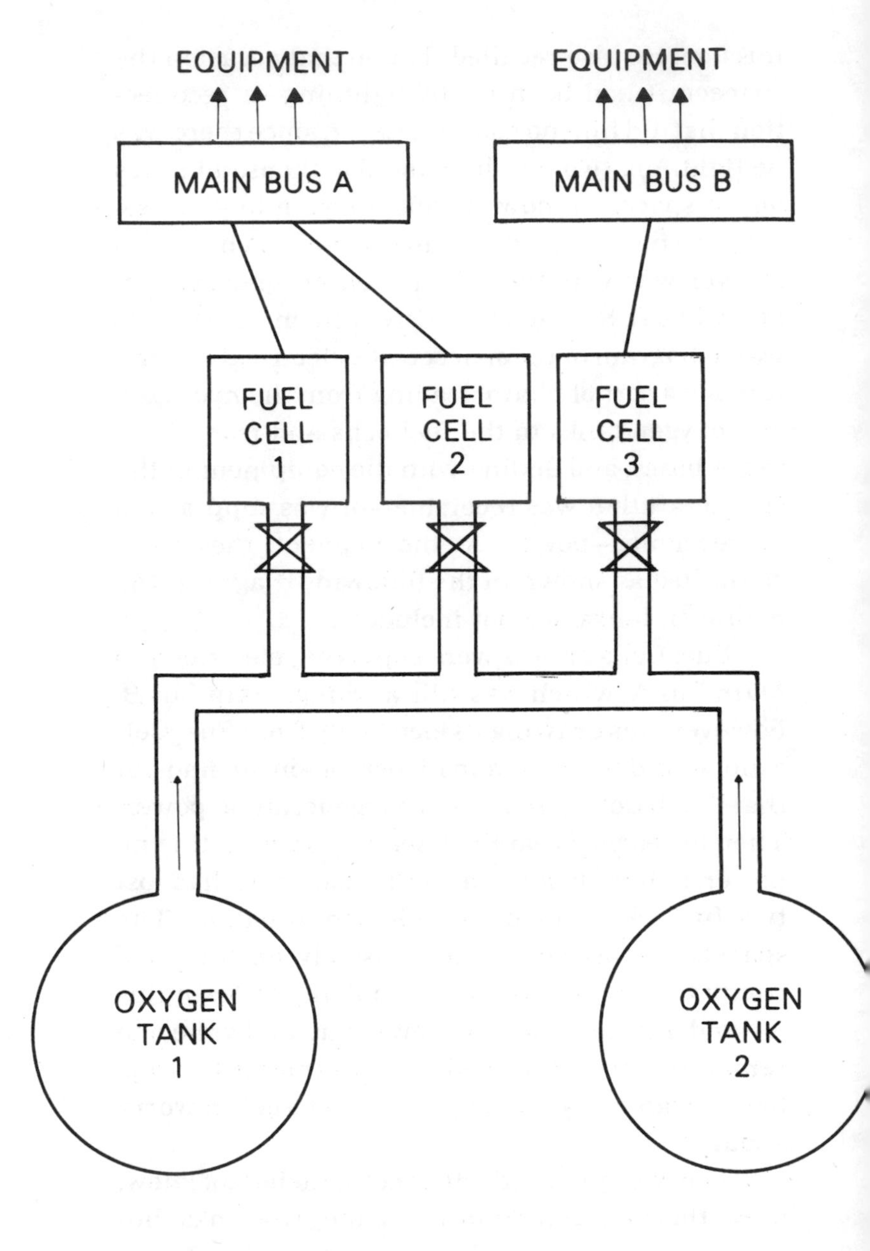
EQUIPMENT
EQUIPMENT
MAIN BUS A
MAIN BUS B
FUEL CELL 1
FUEL CELL 2
FUEL CELL 3
OXYGEN TANK 1
OXYGEN TANK 2
=REACTANT VALVE

since the good cell drew its oxygen from them as well, he didn't give the tanks much thought. There was little reason to, for although the two tanks were not, strictly speaking, redundant—they shared the system of pipes leading to the fuel cells—there were so many safety valves that they might as well have been. They were separated by valves that insured that the oxygen would flow only *out* of them, and, as a further safeguard, each fuel cell could be cut off from the oxygen by a valve of its own, called reactant valves. The protection afforded by all these valves was critical, because in addition to fuelling the electrical system and producing water, the two tanks provided all the command module's oxygen for breathing. (In the cabin, there was a small emergency supply in a tank called the surge tank, and in three one-pound bottles, but this had to be conserved for breathing when the command module plunged alone through the earth's atmosphere.)

The implication of two dead fuel cells was so staggering in itself that Liebergot couldn't bring himself to believe that such a state of affairs was possible. At NASA, backups don't fail. Liebergot was encouraged in his disbelief by the flight controllers' operating procedure, which required them to make presumptions against such failures—partly because of the admittedly imperfect quality of telemetry. According to the standard procedure, before Liebergot could think about such a thing as the oxygen he had to make sure that the fuel cells really were dead. Perhaps there had been an instrumentation failure, or perhaps there was some other simple explanation. Until he was certain, Liebergot didn't even report the loss of the cells. "You can't alarm the crew unnecessarily—you'll look like a big ass unless you're sure," he later said.

One simple explanation that occurred to Liebergot was that the jolt or the bang, whatever it was, at eight minutes past nine had disconnected the two fuel cells from the two buses. He therefore suggested to Kranz that Swigert check on whether the cells were in fact hooked up to the buses. Any connecting or disconnecting that the astronauts did was by means of switches at their consoles; the switches for the cells and buses were in front of Haise but within reach of Swigert, and Swigert now flicked them down and back. (Flight controllers call such flicking "cycling the switch.") Swigert reported no change in the electricity level of the buses. That meant that the trouble did not lie in anything as simple as broken connections. Liebergot couldn't make any sense out of it. He wished he could be almost anywhere else. He couldn't, of course, because, among other things, Lousma kept asking Kranz if there were any more recommendations he could pass on to the astronauts, and Kranz kept asking Liebergot. Liebergot felt rather cornered. He was the one on the spot, and all the electrical engineers in the world couldn't help him. At length, because it was possible to switch circuits among the fuel cells and buses, he suggested that the astronauts switch the lines from the two dead cells so that each fed the other bus. That way, Liebergot figured, something might develop to give him a clearer picture of what was going on. Also, fuel cells, like flashlight batteries, sometimes worked better if they were changed around. Kranz, however, refused to go along with the suggestion. He had to be cautious, because nobody knew what was wrong and he didn't want to do anything that might make matters worse. At the moment, all the power in the spacecraft was coming from Main Bus A, and he didn't want to risk disturbing it. For the time being, Kranz planned to be very deliberate and very methodical about authorizing any changes. Less than seven minutes had passed since the bang.

The only other idea that Liebergot could come up with just then was that in order to keep up the power in the good bus—which was continuing to drop—the astronauts might augment it by feeding into it electricity from one of the storage batteries that were supposed to supply power to the spacecraft during its reëntry through the earth's atmosphere. There was, it seemed, very little choice; in fact, as Liebergot made the suggestion, he could see on his telemetry screen that the astronauts were already hooking the battery to the bus.

Although the brunt of the difficulties fell on the EECOM, all the other flight controllers were having trouble, too. Their voices over the loop were almost unnaturally calm, but one of them said later that he could tell from their tones that "a lot of stomachs were turning over." At the moment of the bang, the spacecraft began pitching and yawing about like a depth-charged submarine. Two identical balls set in the dashboard, one in front of Lovell and the other in front of Swigert, appeared to spin erratically. They were the flight-director attitude indicators, or F.D.A.I.s—sort of three-dimensional compasses that showed which way the spacecraft was pointing. Appearances to the contrary, the balls were actually still, as the compass card in a ship's binnacle is; it was the spacecraft that was doing the turning. The guidance computer in the spacecraft, which normally held it steady by automatically firing sixteen small thruster rockets outside the service module whenever necessary to correct its attitude, had been unable to stop the wobbling. Now Lovell was trying to do so by firing the thrusters manually, using a pistol-grip hand control at the end of the armrest on his couch. He wasn't having much luck, for the spacecraft kept buffeting and yawing as if something were venting from it and imparting an unwanted thrust. Part of Lovell's difficulties stemmed from the fact that the sixteen thrusters operated electrically; each of the main buses supplied current to fire half of them, and the eight thrusters dependent on Main Bus B weren't working. The good bus couldn't accommodate all sixteen, so Buck Willoughby, the Guidance and Navigation Control Officer,—a tall ex-Marine flier from Colorado who sat on Liebergot's right—had to figure out which were the best thrusters to keep operating; he had to make sure that one thruster was working in each direction for the three motions of the spacecraft—up and down, left and right, and roll.

The GNC had a special interest in getting Lovell to steady the spacecraft, for once it had been brought to the right attitude Lovell could set up the gentle roll—the passive thermal-control roll—that kept it turning once every twenty minutes, so that the sun would heat it evenly on all sides. The delicate electronic instruments for guidance and navigation, which were the GNC's responsibility, and some of the spacecraft's propulsion systems were especially sensitive to extremes of temperature, and without the regular thermal roll the part of the spacecraft left facing the sun could get as hot as two hundred and fifty degrees, and the part left facing away could get as cold as absolute zero. However, even after Lovell had plugged the thrusters into the buses in the way the GNC thought best, he still couldn't control the spacecraft's attitude.

The wobbling was causing other problems as well, for if the spacecraft should happen to roll into certain attitudes the guidance system would lock. The heart of the guidance system was the inertial-measurement unit, a spherical structure in the lower equipment bay, at the foot of the center couch, containing the guidance platform. This was a small metal block that swung freely on three gimbals (like those that keep a ship's compass level), so that a set of gyroscopes could maintain it in the same position in relation to the stars regardless of the attitude of the spacecraft. Its attitude was electrically relayed to the ship's guidance computer. The platform had been aligned with certain stars before launch, and it was still aligned with them, for whenever the spacecraft rolled, pitched, or yawed the spinning gyroscopes adjusted the gimbals to keep the platform true. The trouble was that if the three gimbals lined up in certain ways, they would lock and the spacecraft would suddenly be without any reference point in space. In effect, the astronauts would be without a compass. Two or three times, the GUIDO broke in on the loop to tell Kranz that the spacecraft was wobbling toward what he called "gimbal lock," and at the warning Lovell would point the spacecraft in another direction as hastily as a helmsman would steer a ship away from a reef.

A spacecraft is such a welter of interdependent elements that any one problem can set off a whole series of other problems. The erratic spinning threatened radio communications, because it was next to impossible to keep the antennas aimed at the earth. After the accident, the INCO, the radio controller, who sat at Kranz's left, had advised shifting from the high-gain antenna—the stick that had been jarred at the time of the bang—to the omnidirectional antenna system, which didn't have to be pointed so precisely. There were four omnidirectional antennas—big scimitars—spotted around the spacecraft, and normally as it rolled the INCO would keep switching on the one that happened to be on the side nearest the earth. However, with the wobbling, there were times when neither the INCO nor the astronauts knew which antenna was facing the earth. Sometimes communications stopped altogether.

The astronauts were so busy avoiding gimbal lock, checking antennas, and transferring thrusters and other equipment from one bus to the other that they didn't have a chance to worry much about exactly what sort of danger they were in, and most of the flight controllers didn't, either. However, one person who had a little time on his hands for worrying was the TELMU, Robert Heselmeyer; because he was a lunar-module man, he was somewhat removed from the situation. Although the LM was powered down, it was still using a little electricity drawn from the command module to warm some of its equipment. Heselmeyer sat in silence as he watched the current being fed to the LM go down and down. When the current stopped altogether, he reported the fact to Kranz. Kranz asked Heselmeyer to get back to him later, because he had enough on his hands at the moment. Heselmeyer continued to worry, for it had crossed his mind that if anything serious happened to the command module the astronauts might have to use the lunar module as a lifeboat. He rummaged around on his console for instructions on such lifeboat procedures.

Liebergot, who was still trying to come up with some ideas for reviving the two apparently dead fuel cells, suggested to Kranz that they both be unhooked from the buses, in the hope of separating bad equipment from good. Kranz, who was still being cautious and didn't want to disturb too many things at once, agreed to disconnect only Fuel Cell 1, which was attached to the good Bus A. Part of his thinking was that by isolating one section of the system and then another it might be possible to pinpoint the trouble spot.

In the meantime, Liebergot had requested Lovell to read to the ground all the gauges having to do with the electrical system, to see if they bore out the information that the ground was receiving. Lovell at last got as far as the pressure gauges for the oxygen tanks. "Our Oxygen No. 2 Tank is reading zero. Did you get that?" he said. Now he floated up out of his seat and pressed his face against the window so that he could look backward toward the service module. He saw a thin sheet of vapor, like a cirrus cloud. "It looks to me that we are venting something," he reported. "We are venting something out into space." Now he understood why he had been unable to steady the spacecraft: venting imparted motion as surely as firing a rocket. Thirteen minutes had passed since the bang. To Lousma, the CAPCOM, this was the most chilling moment of the flight. Although Lovell's voice was calm, what he was saying was as alarming as if a ship's captain had reported seawater rushing in through the hull—only the astronauts wouldn't be able to jump into anything as hospitable as even an Arctic sea.

Lovell had no doubt that what he had seen venting was a gas, for in space liquids form hard nuggets, not thin sheets of cloud. And he had a pretty good idea that the gas was oxygen, for he now noticed that not only was the pressure in Oxygen Tank No. 2 at zero but the pressure in Tank No. 1 was dropping as well; it was oxygen from this tank that was leaking into space at the moment. Lovell felt that it was just a matter of time before the command module itself would go dead.

The other flight controllers, unlike the CAPCOM, so far had no such forebodings. As the disaster unfolded step by step, they continually seemed to be left incredulous, one step behind. One reason for their incredulity was that they were missing a key fact: they did not know that the original tank failure had been a *violent* one. Liebergot was thinking in terms of a gentle leak, and he did not suspect that the rupture of Tank No. 2 had been explosive; what had actually happened was that it had ripped out pipes and valves between the two tanks, in an area called the manifold, where several pipes joined, also causing the oxygen in Tank No. 1 to slowly dissipate. And there was something else the flight controllers didn't know: the jolt had shut the reactant valves on Fuel Cells 1 and 3—the immediate cause of the power failure, as the valves cut off the flow of oxygen to the two cells. The command and service modules' electricity would last only as long as oxygen from Tank No. 1 continued to reach Fuel Cell 2. As the oxygen in the service module was what the astronauts were breathing, it had already crossed Liebergot's mind that the astronauts could be what he called "belly up" in a matter of hours. He couldn't quite believe it, though, because of his trust in the soundness of Tank No. 1. Kranz gave a short talk to the flight controllers over the loop, urging them to keep cool; guessing would just make matters worse, he said. Optimistically, Liebergot checked to make sure that there was no instrumentation problem.

The astronauts, however, by now had no illusions. Haise said later, "The ground may not have believed what it was seeing, but we did. It's like blowing a fuse in a house—the loss is a lot more real if you're in it. Things turn off. We believed that the oxygen situation was disastrous, because we could *see it venting.* The ground may have been hoping there was an instrumentation problem, but on our gauges we could see that the pressure was gone in one tank and going down in the other, and it doesn't take you long to figure out what happened."

The chief reason the flight controllers didn't tumble to the seriousness of the oxygen problem was that the correct answer was also the unthinkable one. In a number of respects, the situation was like the sinking of the *Titanic,* another craft that was admired as a nation's greatest technical achievement. The ship had reputedly been unsinkable, because its hull was divided into redundant watertight compartments, but the collision with an iceberg sliced open too many of them, and it sank. There were obvious differences between the two incidents—Apollo 13, for example, carried its own iceberg within itself. A remark made later by a NASA engineer was strongly reminiscent of the worldwide reaction to the earlier accident: "Nobody thought the spacecraft would lose two fuel cells and two oxygen tanks. It couldn't *happen.*" Swigert himself wrote afterward, "If somebody had thrown that at us in the simulator, we'd have said, 'Come on, you're not being realistic.' "

For some time, the flight controllers, in their pursuit of solutions, attempted to go in two directions at once: they tried to save the moon-landing mission while simultaneously preparing for the worst. About four minutes after the astronauts reported the venting, Liebergot suggested to Kranz that the crew start powering down the command module, to put less strain on the surviving bus, which was continuing to lose power. Liebergot told Kranz he wanted the astronauts to work their way through the first half of page 5 of what was called the Emergency Power-Down Checklist, which told how to turn off equipment in the proper sequence so that an instrument wasn't switched off before another one that depended on it. For the moment, the procedure should ease the power crisis. In the spacecraft, the astronauts, to find the proper checklist, had to riffle through twenty pounds of instruction sheets before they got the right ones. Kranz still had no intention of giving up the ship; he made sure no equipment was turned off that would preclude landing on the moon—a possibility he had by no means abandoned. Like Kranz, Liebergot was hoping there was still some way of saving the mission, and as he went about selecting more equipment for the astronauts to turn off he was thinking, in another part of his mind, of possible ways to get more oxygen out of Tank No. 1, so that they could power up again.

The power shortage in the command module was now well beyond the help of the reëntry battery, so Liebergot ordered the astronauts to disconnect it. They were reluctant to do so as the battery was providing a good deal of the current they were able to draw from the good bus, but Liebergot insisted. He knew that as the power-down continued, there would be less need for the battery anyway. He had suddenly begun to fear that nothing could be done—that it was just a matter of time before the command module lost all its power. In that case, the astronauts would have to use the LM as a lifeboat to bring them back to earth. Then, just before they hit the atmosphere, they would have to abandon the LM and find some way to fly the dead command module, which was the only part of the Apollo spacecraft with a heat shield capable of withstanding the high temperature of reëntry. If they were to do that, they would need every ampere of electricity in the reëntry batteries. For the same reason, Liebergot urged that the astronauts immediately isolate the supply of oxygen in the surge tank, which was normally connected with the service-module supply. Kranz, who was still thinking in terms of conserving the oxygen in Tank No. 1, wanted to know why Liebergot was ordering the change, and the EECOM replied that he was now more worried about conserving the reëntry oxygen. Kranz saw what he was driving at.

As is bound to happen on any ship with a bad leak, a certain amount of confusion arose. Once, the astronauts turned off a switch that incidentally cut out the gauges for the oxygen tanks; Liebergot quickly got them to turn it back on again. Another time, they failed to turn off a part of the guidance system that the ground had asked them to switch off, and no one discovered the omission until the spacecraft began jerking about. Consequently, Kranz requested the astronauts to read back to the ground the dials of all the two hundred and fifty instrument gauges on two of their dashboard panels, for it was imperative that the flight controllers who would be figuring out how to bring the dead command module back through the atmosphere know exactly what the situation was. The astronauts began their reading-back at 9:59 P.M., and it went on for ten minutes.

As additional precautions, Kranz requested that a two-hundred-foot radio antenna (called a deep-space dish) in Australia be added to the global network tracking and communicating with the spacecraft, and that additional computers at the Goddard Space Flight Center in Maryland be what he called "cranked up"—made ready for use. He also telephoned the Real Time Computer Complex on the ground floor of the Operations Wing to ask that an additional big I.B.M. computer be brought onto the line. The computer complex gets its name from the fact that it processes data in "real time," the flight controllers' term for instantaneously. And since the average Apollo flight transmits fifty-five million bits of data, this takes some doing. Most NASA engineers believe that it was the United States' superiority in computers, more than anything else, that gave this country the ability to land men on the moon when it did. The figuring for the first Soviet manned orbital flights may well have been done by teams of men using desk calculators. The technicians in the R.T.C.C. have consoles much like the ones in the Control Room upstairs, and, indeed, many of the computer technicians are the counterparts of—or, as they put it, they "interface with"—the flight controllers. (The interfacing is done over an intercom.) Through a glass partition, the technicians overlook a brightly lit room filled with computers; each computer consists of a couple of dozen cabinets arranged in a rectangle, like refrigerators in a showroom, and, like refrigerators, they need to be kept cool for maximum efficiency. There are four big computers, each capable of handling a flight to the moon, and a fifth, smaller one, used for simulations.

During the non-critical periods of a lunar mission—the sleep periods and the trans-lunar and trans-earth coasts—all the work was ordinarily done by just one computer, designated the Mission Operations Computer, while a second, designated the Dynamic Standby Computer, was hooked up during critical periods—launch, moon-landing, and reëntry—and served as a backup, receiving all the telemetry information from the spacecraft that the main computer was getting, in case it suddenly had to take over the running of the mission. On the night of the accident, the standby computer was of course receiving no data from the spacecraft. In fact, like the third and fourth big computers in the room, it had been farmed out for totally unrelated work, because NASA was having budgetary problems and computer time is very expensive. After all, there had been only a few times in the past when anything had gone wrong with the primary computer necessitating a switch to the backup, and, besides, if an emergency should arise the primary computer could load all the data about the flight into the backup in twenty milliseconds, which was a very short time indeed. Of course, if the primary computer *was* incapacitated, it might not be able to load the backup at all. Now Kranz wanted to rectify the situation; he had had enough trouble with backups that night.

As Kranz and Liebergot went about their preparations for the worst, they clung to their faith in the spacecraft's recuperative powers. Kranz said later, "We were still hoping to come up with the right configuration of tanks, fuel cells, and buses, and fly out of the woods with the oxygen in Tank No. 1." He held on to this hope in spite of the fact that the pressure in the tank had dropped from nine hundred pounds per square inch, which was normal, to only three hundred. Liebergot at last recommended that the heaters and the fans for

Tank No. 1 be turned on, so that the heat could increase the pressure and the flow of oxygen into the fuel cell. Liebergot studied his telemetry screen. He saw a sudden jump in the amount of current leaving the good bus, and knew that the fans and heaters were on. He did not, however, see any compensating jump in the oxygen pressure; in fact, it was continuing to drop at about the same rate. If anything, the drop was a little faster; the heat almost certainly accelerated the leak. As he watched, the pressure dropped three pounds per square inch, and this made Liebergot pessimistic once more. He said to Kranz, "You'd better think about getting into the lunar module and using the LM systems."

Forty-seven minutes had passed since the accident. Kranz now turned his attention to the lunar module. He told Heselmeyer, the TELMU, to get some men started figuring out the minimum power-up of the LM needed to sustain life. Kranz wasn't yet through with the command module, however. He asked Liebergot if he had any more suggestions for restoring the oxygen pressure. Liebergot had not, but he canvassed some of his assistants—each of the flight controllers was backed up by a team of experts in the Staff Support Rooms, just outside the Control Room—and one of them came up with the idea that the oxygen leak might not be in the tanks at all but, rather, in one of the fuel cells. Fuel Cell 3 seemed like a good candidate, since it had been the first to fail. Liebergot immediately called Kranz and suggested that the astronauts close that cell's reactant valve, and so cut it off from the tanks. If the assistant was right, the leak would stop. It was the last hope for a simple way out. The reason nobody had suggested shutting the reactant valve earlier was that it was an irreversible act: the fuel cell would be permanently out of service, and there would be no possibility of landing on the moon. Kranz had at last stopped thinking about a lunar landing. However, Haise, on whose side of the spacecraft the reactant-valve switches were, had no intention of taking an irrevocable step that would abort a four-hundred-million-dollar mission all by himself—his own pessimistic evaluation of the situation to the contrary—and, accordingly, he asked the CAPCOM to repeat the order, to make sure there was no mistake. It would have saved him some uncertain moments if he had known the valve was already shut—and had been since the bang—but then nobody knew that. Liebergot gazed at his electronic screen to see what effect isolating the fuel cell would have on the oxygen leak. It would take a while to find out. In fact, it took longer than he expected, because Haise, reluctant to perform this irrevocable act, had to be given the instruction a third time before he flicked the switch for the valve. After an interval, Liebergot was able to determine that the oxygen was still going down at the same rate. At last, Kranz and Liebergot were face to face with the possibility that they were dealing not with a leak but with some sort of explosion that had knocked out the entire oxygen system. At last, they were compelled to accept the fact they had resisted for so long—that their main craft's redundancy had failed them. Fortunately, that craft was better prepared than most sinking vessels. NASA had provided it with the ultimate in redundancy—a whole other craft.

Some time later, when Kranz was asked whether he had ever feared for the lives of the astronauts, he replied, "Yes and no." Then he added, "I guess the answer is no, because I have worked with the lunar module more than any other Flight Director. I had the utmost confidence in the LM and in the flight controllers. I knew that the life-support system was good, the communications were good, and the guidance system was good, and that it could make long rocket burns. I was sure it would prove to be a reliable spacecraft."

Once again, Kranz was more confident than his flight controllers, for Heselmeyer, who, as TELMU, was in charge of the LM's electrical and environmental system, was not at all sure that the LM, which was designed to support two men for about two days, had enough consumables to support all three astronauts during the entire journey back to earth. "Consumables" was NASA's term for the fuel, water, and power that the spacecraft consumed and the oxygen and water that the astronauts consumed. A less euphemistic term might have been "essentials." It looked as though the TELMU was going to have to make the LM's consumables last *three* men for *four* days, because the Retrofire Officer, Bobby Spencer, who was in charge of emergency-return plans, and who had sent Kranz the note just before the accident that the spacecraft was close to the point where it could no longer return directly to earth, was talking now more and more in terms of going home around the moon. In the interval since he had sent the note, the spacecraft had travelled about three thousand miles. The moon was now only forty-five thousand miles away, and the earth five times that distance behind the spacecraft. When the spacecraft left the earth, it had been travelling at about twenty thousand miles an hour, to escape the planet's gravity, but as it rose higher above the earth its speed slowed—like that of a tossed-up ball near the top of its rise—to a little over two thousand miles an hour. Now it was about to pick up speed again as it fell toward the moon. If the direction couldn't be reversed, the TELMU was going to have to stretch the LM's human consumables threefold and its power twofold.

Like everyone else at NASA, those who were concerned with the LM's consumables had never dreamed that a command module would go completely dead, let alone do so at the point where it had the farthest possible distance to travel back to earth. (The only worse time for the accident to have occurred would have been when the astronauts were on the moon, in which case the command module would have been without its lifeboat.) One flight controller later made much the same observation as Swigert: "This particular situation was so far down the line that it was exceedingly unlikely, and if anyone had asked us to simulate it ahead of time we would all have said he was being unrealistic." For much the same reason, there hadn't been enough lifeboats aboard the *Titanic,* and the passengers had had no boat drill and so didn't know how to use them. NASA engineers had been relying for some time on the fact that they could use the LM as a lifeboat in an emergency, but they had not paid much attention to what they called "the lifeboat mode." The few simulations of such a problem they had run had been short-term affairs, because they had always assumed that if anything should happen to the command module the astronauts could repair it and be back inside it in a few hours. Aside from some tests a year earlier at the time of Apollo 9, which didn't quite cover the present situation, no one had ever experimented to see how long the LM could keep men alive—the first thing one needs to know about a lifeboat.

In fact, the lifeboat mode was considered such an unlikely eventuality that ordinarily during the long coast to the moon, when the lunar module was inactive, the LM's consoles in the Control Room were not manned. By a lucky chance, a TELMU—Heselmeyer—happened to be present at the time of the bang (he had come in to supervise what was known as a "LM housekeeping"), and a few minutes later Kranz had come on the loop to say that he wanted the LM consoles manned around the clock from then on. Heselmeyer's initial estimate of the LM's consumables was so chilling that he called the RETRO and tried to talk him into doing a direct abort—aiming the spacecraft toward the earth and blasting home with the big rocket in the service module. Theoretically, it could still be done. A direct abort could take as little as a day and a half, which would be much better from the TELMU's point of view than a trip around the moon, which he estimated would take about four days.

This conversation took place over a secondary loop—one of several telephone hookups that the flight controllers could talk on among themselves without interrupting the main loop of the Flight Director. The Retrofire Officers were not happy about the TELMU's request. There were two RETROS now, for Spencer, the one on duty, had been joined by the Lead Retrofire Officer, Charles Deiterich, who would have the chief responsibility for planning the route back. Deiterich—a tall, laconic man with a droopy mustache, who was a graduate of the University of St. Thomas in Houston and had joined NASA in 1964—told Heselmeyer that at the moment the spacecraft was so close to the moon's gravitational pull that in order to blast the command module straight home the lunar module would have to be jettisoned, because the main rocket was not strong enough to reverse the direction of the entire spacecraft. Liebergot, who was listening in on this conversation, now said that he could not approve any plan that meant getting rid of the LM. Deiterich added another argument against the direct abort: the spacecraft was getting very close to the point at which a maximum burn of the main service-module rocket not only wouldn't reverse the spacecraft's direction but would simply slow it so that it crashed into the moon. And—assuming that point hadn't been reached quite yet—if for any reason the big rocket couldn't be brought up to full thrust, the spacecraft would crash into the moon anyway. As Deiterich listened to the conversation from the spacecraft, he wondered if the rocket could be fired at all, in view of the electrical failure. Heselmeyer went back to considering ways of stretching the LM's consumables over the estimated four days. To get some help, he put in a call to the Spacecraft Analysis Team (SPAN), a group of engineers in a nearby Staff Support Room who were in constant touch with the prime contractors of the spacecraft—for the LM, the Grumman Aerospace Corporation in Bethpage, New York, and for the command and service modules the North American Rockwell Corporation in Downey, California. At Grumman and North American, there were subsidiary SPAN groups in touch with a number of subcontractors; Heselmeyer's questions were referred to, among others, technicians at the Hamilton Standard Division of the United Aircraft Corporation at Windsor Locks, Connecticut, the subcontractor for the LM's environmental-control subsystem. (Through this cross-country network, the SPAN engineers would get the flight controllers answers to about a hundred and fifty questions; the wires were already humming between the SPAN room in Houston, the SPAN room at North American, and one in Boulder,

Colorado, at the Beech Aircraft Corporation, the subcontractor for the hydrogen and oxygen tanks in the service module.)

Before altogether ruling out the direct abort, Deiterich got the technicians downstairs in the R.T.C.C. to run several possible direct-abort trajectories through their computers, and he passed on a list of seven potential landing sites to the Recovery Operation Team, which was in charge of picking up the astronauts when they splashed down. The Recovery Officers, who were in a glassed-off room to the right rear of the Control Room, were even worse prepared for the emergency than the TELMUS. Because of the success of previous spaceflights, NASA had felt justified in gradually lessening the number of rescue ships. During the Mercury and Gemini programs, when astronauts were orbiting the earth, there had been twelve or so recovery ships stationed around the world at such intervals that no matter where the spacecraft came down there would be a ship within a few hundred miles. However, as spacecraft began going all the way to the moon the number of ships had been reduced, on the theory that in the event of trouble there was more chance to guide a spacecraft to a specific landing site, fewer degrees of latitude on which to spread out the ships, and more time to dispatch a recovery force to meet it; airplanes could drop frogmen to open the capsule's hatch at any point on earth in a matter of hours. No one seriously believed that a crippled spacecraft falling from the moon might need more than frogmen to meet it. By the time Apollo 11 made the first landing on the moon, the number of recovery sites had been cut down to four—what Recovery Officers called the Mid-Pacific Line, the West-Pacific Line, the Atlantic Ocean Line, and the Indian Ocean Line. By the time of Apollo 13, all four of these stations still existed in the minds of Recovery Officers, but only one of them, the Mid-Pacific Line, had any ships at it. When the Recovery Officers were told that they might have to have a rescue force at any one of seven places within as little as thirty-six hours, they swung into action. First, they called the Department of Defense to see if any United States Navy recovery ships happened to be near any of the targets; they also surveyed merchant shipping around the world for what they called "ships of opportunity"—a hair-raising idea at NASA, where nothing is supposed to be left to chance. Twelve countries volunteered ships. The nearest ship of opportunity to the Atlantic Ocean Line turned out to be a carrier, the U.S.S. *America,* which had just left Puerto Rico and couldn't get to the site until twenty-four hours after a splashdown there.

An hour and nine minutes after the bang, the White Team handed over control of the spacecraft to the Black Team. Some minutes before this, Kranz had announced over the loop that the TELMUS and Recovery Officers who were working on plans for a direct abort back to earth could forget it—they did not know what shape the main rocket in the service module was in, and they had more confidence in the LM's main rocket, the Descent Propulsion System. The spacecraft would be going around the moon. There was relief in the Recovery Room, if nowhere else.

Kranz had decided to go ahead with the change of shift despite the crisis, because he felt that he had gone as far as he could and that it was time for a fresh team. Flight controllers take great pride in handing over to a new team on time, regardless of the circumstances. As Chief of the Flight Control Division, Kranz had spent a lot of time training the controllers to achieve what he called "uniformity of decision" from one shift to the next, so that handovers would be smooth, and so that Mission Control would speak in a consistent way to the astronauts. Each type of flight controller had its leader—a Lead RETRO, a Lead GUIDO, a Lead EECOM, and so forth—who reported directly to Kranz, and Kranz saw to it that the men who performed the same job on different teams shared offices, so that the four RETROs, for example, would come to know each other and each other's way of doing things extremely well. For the same reason, Kranz encouraged all his flight controllers to be well acquainted, and he believed them to be an even closer group than the astronauts (against whom they sometimes played touch football), for whereas the astronauts were divided into a number of three-man Apollo teams, the flight controllers—more than a hundred of them, including assistants—flew all missions together.

One of the most difficult arts for flight controllers to master, Kranz felt, was the ability to turn a problem over to someone else when a fresh approach was called for. As Liebergot handed over his console to the new EECOM, he felt his throat tighten up so that he could barely talk—a reaction he says he is likely to have immediately after an emergency. He though he was lucky that his voice hadn't cracked before. He felt like a Jonah, however, as he walked away from his console, reflecting that major electrical problems always seemed to turn up when he was around: he had been on duty when Apollo 10 lost a fuel cell, and he had been present when Apollo 12 was hit by lightning. He would have felt even more like a Jonah if he had known that the routine cryogenic stir he ordered had triggered the tank failure—which, of course, was in no way his fault. After he found out, he was heard to remark that if he had just let things be the oxygen tank would have blown up on the next EECOM.

Liebergot followed Kranz and the rest of the White Team down to a meeting room on the second floor, one of the Staff Support Rooms. The second floor, whose Control Room and Support Rooms were largely unused, was a duplicate of the floor above; the spacecraft wasn't the only part of NASA built for redundancy. There Kranz and his men spent the rest of the night going over the events of the last hour to see what more they could learn from them, and planning what to do next to terminate the mission safely. (A Public Affairs Officer at the time referred to this task as "looking toward an alternate mission"—as though the astronauts had just taken it into their heads to go somewhere other than to the moon.) The Black Team picked up so smoothly where the White Team had left off that the astronauts in the spacecraft were unaware of the change of shift. The new Flight Director, Glynn Lunney, continued Kranz's efforts to power down the command module while simultaneously seeking ways to restore the pressure in the good oxygen tank. On the admittedly unlikely theory that the leak might be in Fuel Cell 1, Lunney ordered its reactant valve closed. (Again, as in the case of Fuel Cell 3, the valve already was shut.) As before, Haise asked to have the order confirmed and reconfirmed. After he had shut the valve and found that it made no difference, Haise decided that the time had come to abandon ship. "Right about then, it was quite apparent to me that it was just a question of time before the command module was going to be dead," he said later. "So I kind of lost interest in my position there and headed for the lunar module."

As Haise moved through the tunnel into the lunar module, he felt as if his world were turning over. One reason was that the two craft were joined together top to top, so that the direction that had been toward the floor in the command module was toward the ceiling in the LM. There were no lights, except for a flashlight he had brought with him. While the inside of the command module was a cone, the interior of the lunar module was a cylinder laid sideways—though almost unrecognizable as such because of consoles and cabinets jutting with sharp angularity from the walls. Since the LM was designed for only two men, its cabin was smaller than the command module's, and where the third astronaut would fit was hard to figure out. The dashboard panels, much smaller in area than those of the command module, had many of the same instruments, among them the two flight-director attitude indicator balls, a red abort light, and a computer keyboard. There were no seats, and each of the astronauts had to stand, gripping the hand controls for the thruster rockets, like a sailor at the wheel of a ship. Two triangular windows were canted downward and to the side, so that the astronauts could look down at the moon's surface as they landed, but the windows were placed just wrong to give Haise a view back along the spacecraft toward the damaged service module.

Haise had to get things running in the LM. He found three checklists for powering it up under various circumstances, but none of them fitted the present situation, because all three were based on the assumption that the LM would be receiving power from the command module. The Flight Director asked the TELMU if *he* had any checklists that would help Haise. After a search of several minutes, the TELMU came upon a set of instructions for starting the LM up on its own batteries. Beyond this, however, the routine checklists, compiled in preparation for a lunar landing and based on the assumption that there was plenty of time for each step, were long and complex; following all the steps on the simplest one would take two hours, and the CAPCOM had just broken the ominous news that there was only about fifteen minutes' worth of power left in the command module. Since there was no checklist that met the particular situation, Haise and the TELMU improvised one. Their training had not taught them to do such a thing directly, but they found they were so familiar with the spacecraft that they could do it very smoothly—a facility that was about all anyone would have to rely on for the next several days. The TELMU jumped back and forth among the different checklists in front of him, dropping an item here and picking one up there. He told Haise to omit powering up the LM's main rocket, which wouldn't be needed for some time, but urged him to get the guidance system started right away, so it would be ready when the coördinate numbers for the guidance-platform alignment were transferred to it from the command module. While the TELMU was juggling all these items in his mind, he received a call from the EECOM, who urged that Haise turn on the LM's cabin oxygen right away, for the command module's supply was about to be cut off.

All these preparations for the worst didn't prevent the Flight Director from taking one final crack at saving the ship. Just as Kranz had looked to Liebergot for encouraging signs, Lunney looked to his EECOM, who similarly kept dashing his hopes. The Flight Director had noticed signs of life in the temperature and pressure gauges for Oxygen Tank

No. 2—the one that had ruptured—and he now asked the EECOM whether it was possible that the tank still contained oxygen.

"Not likely, Flight," the EECOM answered.

A few minutes later, the Flight Director took a new tack: "EECOM, are you satisfied that both of these oxygen tanks are going down and we're past helping them? I'm just trying to be sure that you're satisfied there is nothing else we can do."

All that the EECOM could think of was to continue to power down the command module.

A few minutes after that, the Flight Director said, "EECOM, let me try one more time. Is it possible that if we got power to Main B we could get Oxygen Tank No. 2 powered up, and up in pressure?"

The EECOM replied, "We don't think that is a possibility, Flight."

Nevertheless, the EECOM suggested that the astronauts turn on the fans in Oxygen Tank No. 2—the same action that had precipitated the bang. It made no difference.

"We've got to get them into the LM, Flight," the EECOM said, and the Flight Director said to the CAPCOM, "Get them going into the LM. We've got to get the oxygen on in the LM."

Lovell went to join Haise in the lunar module, leaving Swigert alone in the command module. Swigert turned off most of the command module's thrusters and the pumps for the fuel cells. Almost the only things that were left on were the cabin lights, the radio, the guidance system, and the heaters and fans inside the remaining oxygen tank. In the Control Room, the Guidance and Navigation Control Officer said to the Flight Director that he hoped the heaters in the command module's guidance system could be left on even after everything else had been turned off. Those heaters had never been switched off during a flight, and if the electronic components of the guidance system became too cold there was no assurance that they would work in approximately four days' time, when they would be needed to guide the command module through the atmosphere. The power for the heaters would have to come from the LM. The Flight Director said that he would see what could be done but that he was sure the TELMU would not want to spare the electricity.

The guidance computer was still on, because there was one more service it could perform before the command module went dead—that of lodestone for the guidance system in the lunar module. The platform alignment had to be transferred from one to the other. In the command module, Swigert read off the gimbal angles of the three gyroscopes—the degrees of roll, pitch, and yaw. The command and lunar modules were not perfectly in line with each other, so these numbers had to be revised. Because Lovell was getting bleary, and it was essential to have the numbers right, he asked the GNC in the Control Room to do the figuring for him. When he at last punched the corrected numbers into the LM's guidance computer, he felt he had passed the first major milestone on the way home. The transfer of the guidance-platform alignment brought life to the LM like a fire in a cold hearth.

It was done not a moment too soon, for Swigert reported another amber caution light in the array shining overhead—the Main Bus A undervolt warning. Quickly, Swigert switched off the last of the thrusters and the guidance computer. He even turned off the small platform heaters, for the GNC had just told the Flight Director he was willing to gamble that the cold would not damage the electronic equipment.

The very last item to be shut off was the reactant valve on the remaining fuel cell. When this had been done, Swigert, as pilot of the command module, told the two other astronauts that getting home would be up to them now. The command module was completely dead. The supreme achievement of American technology had broken down utterly. All that was left was a spacecraft whose very complexity made it harder to handle, plus a group of flight controllers and three astronauts who were themselves products of the vast bureaucratic machine that had produced the malfunctioning spacecraft. On the face of it, this might appear to have made it all the more difficult for them to get outside the situation and impose their will on the wayward spacecraft. However, the accident had also demolished most of the technological appurtenances, such as checklists and flight plans, which substitute a sort of delayed time for immediacy, and also much of the automatic equipment aboard the spacecraft which performed tasks that earlier mariners would have performed for themselves. Now the flight controllers and the astronauts were no different from any other sailors facing disaster at sea. They would do a lot better by themselves than their elaborate paraphernalia had done by them.

About three hours had passed since the bang.

AROUND

TO THE PEOPLE FLYING APOLLO 13, THE FLIGHT AFTER the accident seemed to break into three distinct periods: first, the time until the closing down of the command module; then the twenty-hour period during which the astronauts rounded the moon; and finally the trip back to earth, culminating in the descent through the atmosphere. The second period was one of deep uncertainty, and during it the flight controllers had to decide in the broadest terms what was to be done.

In the Staff Support Room on the second floor, the members of the White Team had swept plastic covers off closed-down consoles in order to turn on the headsets and television screens that gave them access to the same communications loops and the same information from the spacecraft available upstairs. Kranz's team would continue to do the long-range planning that the Black Team was too busy to do. Astronauts, who like to think that they are the ones who fly spacecraft, afterward called Apollo 13 "a ground show." The flight controllers themselves were to call it "a RETRO's mission," for it was the Retrofire Officer, assisted by the Flight Dynamics Officer, who was in charge of the trajectory home. RETROS and FIDOS, who sit next to each other in the Trench, work together so closely that it is sometimes hard to tell them apart. On the way to the moon, when the FIDO is planning the outward-bound trajectory, the RETRO is supposed always to have a plan ready to bring the spacecraft home in the event of trouble, but on the way home the RETRO takes charge of the trajectory and the FIDO helps him out by keeping track of where the spacecraft is. At the moment, the Apollo 13 spacecraft was on a trajectory that would carry it around the moon and swing it back toward the earth, which it would approach after more than four days, but because the spacecraft had left its free-return trajectory it would miss the earth by some forty thousand miles.

Deiterich, the Lead Retrofire Officer, ran through some of the possible alternatives for correcting the course. The first was that the astronauts could make a small burn with the lunar-module rocket to alter the present course just enough to put the spacecraft back on a free-return trajectory, so that after rounding the moon it would hit the earth. However, the TELMUS instantly objected—the Lead TELMU, William Peters, now joined the meeting—

because the burn would not speed up the spacecraft by much and they were averse to any plan that would leave the astronauts dependent on the LM's consumables for almost four days. The consumables picture had brightened somewhat; the Spacecraft Analysis Team had reported back that there was more than enough oxygen in the LM to last the approximately four days it would take to get back to earth. However, electricity and water were still in doubt.

Deiterich moved on to his next proposal, which was to do nothing now but, rather, wait until about eighteen hours had passed and the spacecraft had rounded the moon, and then blast the LM's rocket for all it was worth. This scheme delighted the TELMUS, because it would get the astronauts home in about two and a half days, but the Control Officers—the Lead CONTROL, Harold Loden, had just arrived at the meeting—wouldn't hear of it. Since each flight controller had his own interest to protect, such discussions (there would be more in the next few days) tended to sound a little like the conflicting voices inside the head of a person with a difficult decision to make. The CONTROLS, who were in charge of the LM's guidance and propulsion systems, argued that this second proposal would mean burning the LM's fuel almost to the point of depletion, leaving no margin of safety for mid-course corrections.

Deiterich's third proposal was a combination of the first two: he suggested making a small burn in the next hour or so to put the spacecraft on a free-return trajectory into the earth's atmosphere (though it would not be precise enough for a landing) and then making a second burn after the spacecraft had rounded the moon, to refine the craft's aim and speed it up. Several of the flight controllers opposed this proposal, because it was complicated; nobody liked the idea of doing two burns when one might suffice. The TELMUs, for instance, said that two burns would require twice as much electricity and water as one—every time a burn was made, equipment had to be powered up and kept cool—and, of course, electricity and water were in short supply. Deiterich, however, preferred the third proposal, because it was the most flexible; it left the option of doing a rocket burn after the spacecraft had rounded the moon, and the burn could be a fast one or a slow one, depending on the circumstances then. Kranz approved, and the plan was adopted. One part of the plan did make everyone feel better, and that was getting the spacecraft back on something approaching a return trajectory right away. Lovell said later that his main concern at this point was to get at least into the earth's atmosphere, because he felt that it would be better for the ship to burn up like a meteor than not to come back at all. Deiterich and the other flight officers in the second-floor Control Room wanted to do the first burn at once, but the men in the spacecraft asked for time. The burn was put off for over an hour—until almost three in the morning.

The Black Team, which was now flying the mission, was glad to have the extra time, too, for a rocket burn is a little like coming about in a sailing ship, in that it demands plenty of preparation. Like coming about, a rocket burn is a change in course. Before the RETRO could plan the burn, therefore, the FIDO had to find out precisely where the spacecraft was—for the venting had changed the craft's speed, knocking it out of the trajectory it was supposed to be following. Radio and radar tracking stations around the globe took readings of the spacecraft's position every ten seconds, but these in themselves did not show exactly where the craft was, or where it was going. The raw data had to be processed by computer, and the technician down in the R.T.C.C. who was doing this was having trouble, because the radar data were what he called "noisy" (the points of information were scattered and ambiguous), so it was difficult for the computer to work out vectors—a vector being a point in space where the spacecraft was known to have been. Whenever the technician was confident that he had a reliable new vector, he signalled the FIDO, who found it on his electronic screen. The succession of vectors constituted the spacecraft's actual trajectory. Tonight, the FIDO felt he needed more than the usual number of vectors before he had what he called "a good hack on the trajectory." The spacecraft, he found, was wandering farther and farther off its original course.

When the FIDO felt he had a good hack on the trajectory, he passed the information on to the RE-

TRO, on his left, and the GUIDO, on his right. The three dynamics engineers in the Trench were so close to the big screens at the front of the Control Room, with their diagrams of the earth and the moon, that they sometimes had the sensation that they were flying the spacecraft themselves. Like pilots, they were a close-knit group, and they lit their cigarettes with matches that had "The Trench" printed on the covers, like those of a private club. They regarded the systems engineers behind them (the TELMUS, the EECOMS, the CONTROLS, and the GNCS) much as pilots regard mechanics, and they referred to the computer engineers downstairs as "electricians."

The men in the Trench never feel more like pilots than during a rocket burn. As soon as the GUIDO got the trajectory information from the FIDO, he punched the numbers into a white keyboard in front of him, preparatory to loading (or, as GUIDOS say, "uplinking") the data into the computer aboard the spacecraft. After the information was received aboard the spacecraft, a confirmation copy of it—a sort of return receipt—popped up on the television screen on the GUIDO's console, and he checked it, then pressed a button ordering the spacecraft computer to accept the data. Meanwhile, the RETRO was writing out the instructions for the burn. On a green sheet of paper lined into boxes called a maneuver PAD—a NASA acronym for "pre-advisory data"—he jotted numbers indicating the exact instant the rocket should be turned on, the length of time the rocket should fire, and the attitude of the spacecraft while it was firing. The CAPCOM read these numbers up to Lovell, who jotted them down on an identical maneuver PAD, read them back for confirmation, and finally punched them into his computer.

The mechanics in the second row were as busy as the pilots in the Trench. The Control Officer, whose responsibility for the lunar module corresponded to that of the GNC for the command module, had to modify a checklist for the powering up of the LM's main rocket—the Descent Propulsion System, which Control Officers call DPS (pronounced "dips"). It was at the opposite end of the combined Apollo spacecraft from the big rocket in the service module, and although it wasn't as powerful as that rocket, it could build up the necessary thrust by being burned for a longer time. There was some question in everyone's mind, though, whether the DPS would work at all, for nobody yet knew the full extent of the disaster—perhaps the DPS was out of commission, too. The rocket swung on gimbals to facilitate landing on the moon and also to allow for shifts in the center of gravity during firing; at the moment, the gimbals were set for the lunar landing. The coming burn would be what Control Officers called a "docked DPS," to denote that the LM was docked with the command and service modules and the DPS would be pushing all of them—a tricky setup, because there was some flexibility in the connection between the LM and the command module, and the LM would not necessarily be exactly lined up with what it was pushing. Just before the burn, the CAPCOM radioed up to tell the astronauts that the nozzle of the DPS rocket was not aimed perfectly for a docked-DPS firing, and the astronauts hastily "trimmed the gimbals"—a little the way a seafarer trims his sails. If the aim hadn't been rectified, the spacecraft could have tumbled during the burn and gone off course. The Control also reminded the astronauts to open up the LM's four landing legs, which were folded up beneath the rocket, in the way of the flames. Then, as the astronauts were making a last-minute check of their control panels, one of them mentioned that a switch that could have jettisoned the bottom half of the LM—the part ordinarily left behind on the moon, which contained the DPS rocket—was on. The CAPCOM told the astronauts to turn off the switch.

At two-forty-three in the morning, five and a half hours after the accident, Lovell's left hand pushed a button. For thirty seconds, Lovell felt himself pressed gently toward the floor—the only physical indication he had that the rocket was firing. A planned minor correction later would aim the craft into a narrow corridor through the earth's atmosphere which would bring the astronauts, in less than four days' time, to the Indian Ocean. The Recovery Officers had never stationed a ship in the Indian Ocean, and they had to act quickly now. The nearest United States Navy ship was the destroyer *Bordelon,* which was cruising off Mauritius. They discovered that the *Bordelon* was one of several Navy ships that had been modified to take a special crane for picking command modules out of the sea. However, the Recovery Officers would have to order the crane flown from Norfolk, Virginia, to Mauritius. They were banking on the fact that the *Bordelon* would get to port, pick up the crane, and be at the splash-down site, seven hundred miles away, before the astronauts could get there.

The rest of the night was largely a holding action. In particular, the TELMUs were trying to hold on to the LM's consumables—a task that recurrently brought them into collision with the other flight controllers. As soon as the rocket burn was over, the TELMUs requested that the astronauts begin powering down equipment—especially the guidance computer, since its gyroscopes, which kept the platform steady, were heavy users of electricity and water. The guidance platform wouldn't be needed until after the astronauts had rounded the moon, some fifteen hours later. Immediately, the GUIDO protested that he didn't want to lose the platform alignment, which had been transferred so carefully from the command module. The TELMUs replied that before the next burn the astronauts could set up a new platform alignment with the spacecraft's sextant by taking sightings on stars. The GUIDO wasn't sure that this was possible; Deiterich, the RETRO, insisted that the platform be kept up all the way through the next burn, too; and, upstairs, Lunney, who often backed the pilots against the more conservative mechanics, agreed, although he permitted the F.D.A.I. balls—the dashboard display for the platform, which also used electricity—to be turned off.

This decision put Peters and Heselmeyer, the two TELMUs downstairs with the White Team, who had been working up what they called a "power profile" for rationing the LM's electricity, in a tough spot. It was true that if all went well the trip home could take considerably less than four days; even so, the TELMUs had to think in terms of what they called "worst-case planning"—a conservative approach incumbent upon anyone rationing consumables. Even more than electricity, the supply of water for cooling the LM's electronic gear was a source of worry to the TELMUs. There was a direct relationship between the two; the more electricity the instruments drew, the more water they needed, so cutting down on power consumption was a way of saving water. The cooling system worked like a car radiator, except that the water was not recycled; all the while the LM was powered up, water slowly steamed off into space, boiling away so gently that, as far as anyone could tell, it had never disturbed a trajectory.

Of all the TELMUs' problems, however, the worst was not conserving a consumable but the reverse: how to get rid of the carbon dioxide that the astronauts were exhaling. Normally, the air in the cabin of either the LM or the command module was constantly passed through pellets of lithium hydroxide (small white pebbles that have an affinity for carbon dioxide and therefore remove it) in a process astronauts and flight controllers call "scrubbing the atmosphere." The pellets were inside canisters that fitted into a sort of ventilating system that sucked dirty air from the cabin and then blew the scrubbed air back. When a canister had absorbed all the carbon dioxide it could hold, it had to be replaced with a fresh one. The trouble was that there weren't enough canisters in the LM to last until the astronauts got home, the command module's ventilating system was turned off, and the logical solution, which was to use the command module's canisters in the lunar module's air-purifying system, wouldn't work, because they didn't fit. The problem had never been considered in plans for "the lifeboat mode." Now the TELMUS turned it over to another group, the Crew Systems Engineers, who were responsible for all such equipment, and who would have two days to solve the problem before the astronauts were asphyxiated.

Up in the spacecraft, the astronauts were worrying about their consumables. Haise feared that the cautiously optimistic reports from the TELMUS were a coverup for bad news. He kept double-checking the consumables to make sure that what he was being told was true. Like the TELMUS, the astronauts were most worried about their water supply, and, without telling the ground, Lovell decided that they would stringently ration what they drank. He set Swigert to work transferring some of the drinking water from a tank in the command module to the lunar module. The command module's initially almost unlimited supply of water had stopped, of course, as soon as hydrogen and oxygen were no longer being combined in the fuel cells. Swigert had a hard time transferring the water, because the hose fittings in the two modules didn't match—another curious design defect in two craft built for mutual assistance. Swigert had to use plastic juice bags to make the transfer from the command module to the LM, and in doing so he sloshed water into his shoes. Later, the SPAN engineers would get the technicians at Hamilton Standard, in Connecticut, to look into the feasibility of using the astronauts' backpacks—which have compartments for water—as pails. It would take two days for Swigert's shoes to dry, because with so much equipment turned off the spacecraft was getting cold. He felt as if he were in a leaky boat. In fact, Swigert, who knew least about the lunar module, was the most worried of the three. He stood by a window in his wet shoes watching the earth recede behind them and had some very deep thoughts about never coming back.

Lovell's and Haise's many tasks didn't give them much time for worrying. If they were to avoid the risk of having one part of the spacecraft become overheated by the sun, they had to regain control of its attitude and then set up the thermal roll. There were still occasional spurts of oxygen from the service module, but the venting was less disruptive now. Lovell took hold of the hand control for the attitude-control thruster jets, which in the LM was on the dashboard by his left hand. He ran into trouble right away, because the designers had never intended the lunar module to control the attitude of the entire Apollo spacecraft. The LM was at one end of the combined craft—a bad spot for handling the two other modules—and its thrusters were too weak to handle easily a mass more than twice its own.

As word of the disaster spread, many astronauts had come tumbling into the Control Center to see what they could do to help. One job they could perform was to man the simulators and test maneuvers that hadn't been tried before. The lunar-module simulator was being run by Charles Duke, the backup LM pilot. The simulator, a replica of the LM's cockpit, was hooked up to the smallest of the five computers in the R.T.C.C., and technicians there had already programmed it with data about the present situation, such as the strength and position of the LM's thrusters in relation to the rest of the spacecraft, and even the random effects of the venting. Duke experimented to see if the LM could wrestle the entire spacecraft more easily by firing its thrusters steadily or in short bursts. Short bursts worked better, and this information was passed on to Lovell.

There was no guarantee that the simulator reproduced the motions of the spacecraft accurately, and, indeed, Lovell was finding he couldn't regain control nearly as easily as Duke had done. Because the F.D.A.I. balls had been turned off, Lovell was without a compass, and the only way he could maneuver was by referring to three separate gauges on the dashboard which gave him the angles of roll, pitch, and yaw; it was about as chancy as lining up the three Bell-Fruit bells in a slot machine. Moreover, he was becoming so fatigued that he couldn't remember whether he should fight the spin by going left or by going right, and once he found himself turning entirely around. To add to his problems, he had been having trouble with communications ever since he had begun using the radio in the LM; it made a continual beeping sound, so that he and the flight controllers could barely hear each other. The trouble was caused by a radio transmitter aboard the third stage of the Saturn rocket that had launched the spacecraft and was now trailing a thousand miles behind it on the way to impact on the moon. The transmitter aboard the rocket booster was beeping so that it could be tracked from the ground, and it was using exactly the same frequency as the LM's own radio. This arrangement saved money on ground equipment, and NASA justified it by pointing out that the astronauts weren't supposed to be flying the LM until after the booster had crashed into the moon. On one occasion, Haise told the CAPCOM that he could barely hear him, and the CAPCOM radioed up some emergency instructions for getting home in case communications were lost altogether. Fortunately, the INCO remembered a trick that cut down on the interference. He got the astronauts to turn off their radio for twenty minutes, and during that interval he broadcast a steady signal to the booster on a slightly different frequency, causing its transmitter to shift frequencies. This incident was not the only one in which flight controllers had to play tricks on the overconfidently designed equipment.

Just before dawn in Houston, the spacecraft's attitude suddenly took a turn for the better. By using the jets the LM normally employs for translation—a way of moving the craft up, down, and sideways rather than turning it—Lovell finally managed to get it stabilized and pointed in the right direction—sideways to its trajectory and perpendicular to a plane drawn through the earth, moon, and sun. (Duke in the simulator had been the one to discover that the translation jets were more efficient for jockeying the entire spacecraft.) The next step was to set the spacecraft rolling, for thermal protection, and to keep it rolling—something that the LM's guidance system was not equipped to do, as a LM did not need to roll on its short hop to the moon. Duke had been working on this problem in the simulator, too, and now the CAPCOM radioed up that Lovell would have to rotate the spacecraft some ninety degrees every hour by hand. The CAPCOM promised to remind him to do this.

Around four in the morning, Lovell sent Haise back to his couch in the command module to get some sleep. The command module, which the astronauts had taken to calling "upstairs," would be the bedroom for the rest of the trip. Haise had last looked at his wristwatch before the accident, seven hours earlier, and he had lost all track of time. For him, the intervening period—in which he and the others were abruptly confronted with an almost insoluble problem in a strange place—had had a dreamlike quality.

At about eight o'clock in the morning, after Kranz and the White Team had also gone off to get some sleep, some forty men gathered in the glass-enclosed gallery for visitors, at the back of the third-floor Control Room. They included Robert R. Gilruth, the Director of the Manned Spacecraft Center; his deputy, Christopher C. Kraft, Jr.; and James A. McDivitt, the Apollo Spacecraft Program Manager. Occasionally, the flight controllers of the Gold Team, which had taken over from the Black Team an hour earlier, glanced over their shoulders to see what was going on. A FIDO who happened by said later that he had never before seen so much NASA brass in one place at one time. The NASA brass was trying to decide which of three possible types of burn to do after the astronauts had rounded the moon. The burn was scheduled for eight-thirty that evening, which would be two hours after pericynthion—the spacecraft's closest approach to the back side of the moon—and hence it was called the PC+2 burn. Pericynthion was the point at which the service-module rocket was normally fired to put a spacecraft into lunar orbit, and, in the event that that rocket failed, the point two hours past pericynthion had always been considered the place to do an emergency burn back to earth, because it ordinarily took two hours to power up the LM rocket. (It was the emergency checklist for this burn that the flight controllers had cribbed from the night before to power up the LM.)

Christopher Kraft, who before he became Deputy Director of the Manned Spacecraft Center had once had Kranz's job as Chief of the Flight Control Division, outlined the alternative burns that could be made at PC+2. The first was to jettison the service module and blast the LM's rocket with everything it had, so that the astronauts would arrive in the Atlantic Ocean a day and a half afterward. Nobody liked this idea any better than Kranz's group had liked a similar proposal the night before. It left virtually no room for error—and out around the moon an error in velocity of a tenth of a foot a second could cause a spacecraft to miss the earth altogether. The reason the RETROS had always achieved astonishing accuracy in splashdowns from the moon was that they could, as they put it, "tweak up" a trajectory anywhere along the line with midcourse corrections, and the fast burn to the Atlantic would leave little fuel for tweaking. Besides, in the Atlantic there were no recovery ships.

Kraft hurried on to the two other alternatives, either of which would avert the emergency landing in the Indian Ocean, where the spacecraft was now headed, and bring the astronauts to the prime landing area in the southwest Pacific—the only spot on earth where there were adequate recovery vessels. One of these alternatives involved a relatively fast burn, which would get the astronauts to the Pacific about a day and a half afterward, and the other involved a slower burn, which would get them there exactly twenty-four hours later than that. The twenty-four-hour difference had to do with the earth's rotation—the spacecraft always descended to its splashdown from perigee, the closest approach on the side of the earth that was away from the moon, and consequently the splashdown point depended on the time the spacecraft reached perigee. (The RETROs like to say, "Don't worry—the Pacific will be there!")

While Kraft spoke, those listening to him could see through the glass behind him the big center screen at the front of the Control Room, where the yellow line representing the trajectory of the Apollo spacecraft was moving closer and closer to the moon. From time to time, they glanced at Dr. Gilruth, the Director of the Manned Spacecraft Center, who had previously been director of the Mercury project, the first American manned-spaceflight program. He was the senior man present, and when NASA people met to make a decision there was no voting; rather, after discussion the top man made the decision.

Kraft threw the meeting open for discussion. The faster of the two burns to the Pacific had an immediate appeal, for everyone shared the fear that something else might go wrong with the spacecraft, in which case the sooner the astronauts got home the better; indeed, this seemed such an obvious choice that some astronauts were already practicing it in the simulators. The fast burn to the

Pacific had one serious drawback, however: just as in the case of the even faster burn to the Atlantic, the astronauts would first have to jettison the service module, because the LM was strong enough only to push itself and the command module to the required velocity. Flight engineers have a natural reluctance to do anything as irrevocable as throwing away one third of a spacecraft. Most important, the service module, fitting snugly over the heat shield—the ceramic bottom of the command module, designed to protect the astronauts from the heat of reëntry through the earth's atmosphere—insulated the shield, and no one knew what effect a prolonged exposure to the cold of space would have on it. The service module was normally jettisoned only half an hour before splashdown, and no one had thought it necessary to test the heat shield's resistance to cold for the length of time it would take a spacecraft to come back from the moon.

Kraft wanted to get the advice of the Lead RETRO and the Lead FIDO, both of whom were then on duty with the Gold Team. Deiterich, who had been up all night, arrived at the glassed-in gallery followed by David Reed, the Lead FIDO—a tall man of twenty-eight with light-brown hair, a graduate of the University of Wyoming, who joined NASA in 1964. Reed was more rested, for he had been at home in bed at the time of the accident, and when he turned on his television set and observed Deiterich and several other dynamics officers already in the Trench he had sensibly taken three aspirins and gone back to bed, on the theory that he would be of more use in the morning if he got some sleep. He had had a restless night anyway, and had come in to work at four o'clock in the morning.

Reed and Deiterich, who would be the ones to work out whatever trajectory the meeting settled on, were both opposed to any burn that required jettisoning the service module. Reed pointed out that if they retained the service module and did the slower burn, they would still have the option later, if anything else went wrong, of jettisoning the service module and making a faster burn. NASA engineers tend to favor any alternative that "keeps the options open." Kraft, who was also in favor of retaining the service module and doing the slower burn, summed up the case for this alternative strongly, and Dr. Gilruth, who had said little during the meeting, nodded assent. The men who had to plan how to bring the crippled spacecraft back through the atmosphere were grateful for the extra twenty-four hours that the slower burn gave them.

Deiterich and Reed went back to their consoles, which were side by side in the Trench. Reed's had a single orange light that flickered constantly, showing that telemetry was being loaded into the computers downstairs, but Deiterich's console had no lights whatever. It had two television screens and six clocks. Under the glareproof glass covering the console's desk there was a map of the earth, centered on the Pacific—the target Deiterich was aiming for. As he began planning the PC+2 burn, he also began to consider some of the problems that would come up at reëntry. The astronauts, back in the command module by then, would have to jettison both the service module and the lunar module before the spacecraft hit the atmosphere, but, with a dead service module, the LM would have to do all the work, including jettisoning itself. Nothing of the sort had ever been tried before. Deiterich had a couple of ideas, which he jotted down.

Many of the flight controllers on duty now would be on duty again three days later during reëntry. Behind Deiterich, at the CAPCOM's console, Joseph Kerwin, who had succeeded Lousma at seven-thirty that morning, was talking with the astronauts in the spacecraft. Kerwin, a trim, clean-cut man, was a commander in the Navy Medical Corps; born in Oak Park, Illinois, in 1932, he graduated from Northwestern University Medical School in 1957 and became an astronaut in 1965. Kerwin was having a hard time hearing the Apollo's crew, because the spacecraft's amplifier had been turned off to save power. He had the volume on his headset turned up so high that hours after he went off duty he was still deafened. In spite of the crackling in his headset, he managed to catch one unexpected statement. Loud and clear, and apropos of nothing in particular, Lovell said, "Joe, I'm afraid this is going to be the last moon mission for a long time." That was not the kind of talk that Lovell's superiors expected to hear from one of their astronauts under any circumstances. (This was the only indiscretion, if that is the word for an honest doubt, an astronaut committed during the whole flight. Lovell's fears did not come true, though Apollo 13 may have been a factor in NASA's pending decision to drop two of the then six remaining Apollo flights. However that may be, some design changes were to be made in the spacecraft before the next mission to prevent another such accident from happening. All wires inside the cryogenic tanks were to be insulated with stainless steel; a third oxygen tank was to be added to the service module, at some remove from the other two tanks; and a battery capable of powering the command module home from any point in its orbit was to be added. Some alterations were also to be made in the Control Room: the EECOM's console would be provided with a better warning system, and the philosophy behind the training simulations would change so that, as Reed would say later, "They can throw anything at us they want, and we won't object.")

When Lovell woke up Haise at about ten o'clock in the morning, he asked him how he had slept. Haise hadn't slept well at all. Shortly afterward, Lovell and Swigert disappeared upstairs into the command-module bedroom. They didn't sleep very well, either. The brilliant sun kept streaming in through the windows as the spacecraft rolled about, making disconcerting stabs of light. At Lovell's suggestion, they pulled the shades on the windows. But without benefit of sunlight the cabin got very cold, and without electrical power it didn't warm up again.

The main business Tuesday afternoon was preparing for the PC+2 burn, which was to take place at eight-thirty that evening. First, the astronauts would have to make sure that the alignment of the guidance platform was still accurate, for the gyroscopes that kept the small metal platform stationary could gradually drift out of line, imparting errors. Ordinarily, an astronaut seeking to check the alignment punched into his computer a request that it find a particular guide star. The computer, using the platform as its reference, swung the spacecraft to the right attitude to bring the star into view, and then the astronaut squinted through a telescope—the Alignment Optical Telescope, or A.O.T.—to see if the star was neatly centered in the telescope's field of vision. If the A.O.T.'s aim was off, he computed the angle of error, which was also the degree of error in the alignment, and punched the correction into the computer. Doing this now was out of the question, because clouds of debris particles from the exploded tank surrounding the spacecraft shone so brightly in the sunlight that they completely obscured the stars. That morning, the Lead GUIDO, Kenneth Russell—a tall, curly-haired man, who sat alongside Deiterich and Reed—had suggested that instead of using the guide stars the astronauts check the platform against the sun, which would be a good deal easier to see in the blizzard of particles. Deiterich had complained that a sun check would not be exact enough; whereas a star is a precise pinpoint of light, the sun's disc is so big that a check based upon it would be accurate only to within two degrees. However, Russell had nothing better to offer, and Deiterich couldn't think of anything better himself, so he agreed to accept the two-degree error.

What had made him hesitate at the time was uncertainty whether the error could be corrected later, because the TELMUs had told him that the guidance platform would have to be turned off immediately after the PC+2 burn and kept off all the way back to earth, and Kranz had indicated that there would be no reprieve this time. However, Reed, the Lead FIDO, had found a way out: he had remembered from Apollo 8 a trick for tweaking up the trajectory on the way back to earth without the platform. An alignment involving the earth's terminator, it was almost as simple as a sailor's using the North Star to steer by, but without it Deiterich would never have been willing to accept the two-degree error now.

As Lovell veered the spacecraft in search of the sun, he muttered that he didn't "have all the confidence in the world in this sun check." When the sun appeared in the window, Lovell, momentarily leaving the controls, squinted through the Alignment Optical Telescope. "I've got it!" he said, but just then the LM lurched and the sun disappeared from view. The problem was that the A.O.T. was fixed rigidly inside the LM's window—it didn't swivel like a simpler telescope in the command module—and consequently Lovell had to aim it by maneuvering the LM, which of course couldn't move with the necessary precision as long as the command and service modules were attached. With Haise's help, he managed to get the sun lined up on the cross hairs long enough to align the platform up to the two-degree limit. Kranz, who was back on duty now along with the White Team, was not satisfied—he hoped they would be able to get a finer alignment on a star a little later, when the spacecraft would be in the moon's shadow and there would be no more glare from the sun.

Though the astronauts had not had much time to think about the moon, they were so close to it now that it overflowed the spacecraft windows, filling the cockpit with cold white light. The light, however, lessened and lessened, for they were moving around to the moon's dark side, and at last the moon and the sun as well suddenly vanished. So did their own dazzling halo of debris. The sky outside was now calibrated with precise pinpoints of light, suitable for a star check. However, whenever the astronauts got one of the guide stars in view, it snuffed out—although their debris no longer glowed in the sunlight, it was still there and made black splotches that obliterated whole constellations. At last Kranz told Lovell to stop chasing after them—he was using up too much thruster fuel.

The spacecraft was going around the moon like a boat rounding a buoy. The nearer to the moon the spacecraft came, the faster it moved; it was travelling at six thousand miles an hour now—three times its speed at the time of the accident.

The earth sank nearer and nearer the moon's horizon, and then it, too, disappeared. The astronauts would be out of touch with Houston for about twenty-five minutes—until they emerged on the other side. The orange light on the upper left of the FIDO's console stopped its constant flickering, and he knew that the telemetry from the spacecraft was no longer reaching the computers downstairs. The flight controllers stood up, stretched, and began talking to each other face to face, without benefit of the loop. Normally, the first passage of a spacecraft behind the moon was a suspenseful time for those waiting on earth, but the Apollo 13 mission had been so suspenseful already that most of the flight controllers regarded the period of radio silence as a breather.

The astronauts regarded it as a breather, too, for the pass behind the moon gave them their only chance to take a close look at it; at pericynthion, they were only a hundred and thirty miles from its surface. Before that, the sun had popped up into the sky again, so that they could see the ground. In the early dawn, the mountains below cast shadows longer than their own height—the moon itself looked dappled and dark—but as the spacecraft hurtled on, coming ever closer, the shadows shortened and the ground became increasingly bright. The inside of the spacecraft became brighter, too, and the astronauts put away the flashlights they had been using. Lovell had circled the moon ten times on the Apollo 8 mission, but Swigert and Haise were seeing it for the first time. Coming so far to see what others had seen before, and better, was anticlimactic, but although they were not the first to see the moon so close, they had the disquieting feeling that they could well be the last, and this gave their observations a compensating urgency. The back of the moon was a jumble of whitish highlands, with here and there a small black *mare* nestled among the hills like an alpine lake. Haise clicked away with his camera at one of the black spots, the Crater Tsiolkovsky, until it was lost again in the folds of the mountains. The photographs proved to be the most detailed ever taken of the area, one of the most interesting on the moon's back side. At pericynthion, Lovell pulled the two other men away from the window, reminding them that they had a burn to do in two hours.

On the ground, Kranz, too, was getting nervous about the PC+2 burn, in part because the FIDO had reported some unexplained changes in the spacecraft's velocity. The changes were all the more perplexing because the venting from the oxygen tanks had almost certainly stopped by now, and this was the only cause of such aberrations Kranz or the FIDO could think of. Of course, any unpredictable last-minute changes in the spacecraft's speed—and hence in its trajectory—would further complicate the planning for an accurate burn. Above all else, Kranz was anxious that nothing go wrong with the PC+2 burn and knock the spacecraft off the return trajectory that everybody had worked so hard the night before to achieve. On the radio, the

CAPCOM reminded Lovell that he should cut short the burn at the first sign of trouble; the burn could be done again any time in the next several hours. There was a new CAPCOM now; Kerwin had passed on the crackling headset to Vance Brand, a thirty-eight-year-old graduate of the University of Colorado, who had been a test pilot with the Lockheed Aircraft Corporation before becoming an astronaut, in 1966. Brand, a stocky man with light hair, helped out in the command-module simulator between shifts as CAPCOM. The idea now was to speed up the spacecraft so that it would arrive at its perigee about nine hours sooner; not only would this bring the astronauts back earlier but it would move the landing site from the Indian Ocean to the southwest Pacific, a distance of some ten thousand miles. If the burn had to be cut short, the astronauts could come down anywhere between the two points, and, accordingly, the RETRO on duty, Bobby Spencer, drew a line between them on a map and passed the map on to the Recovery Officers, who would have to be prepared to rescue the astronauts anywhere along it.

The Recovery Officers, who now had to compile a list of all shipping within striking distance of the line, were already nervous, because their meteorologists had announced that a hurricane—Tropical Storm Helen—was heading for approximately the same spot in the Pacific as the astronauts. The Recovery Officers suggested that the astronauts land somewhere else on that longitude—a little east or west. Deiterich, who had been up now for almost twenty-four hours, and who had the plans for the burn all set, strode into the Recovery Room and, as he put it later, "really hounded those guys until I got them to admit that they didn't have enough of a handle on the weather to say *what* would happen in two days' time." RETROS sometimes were as tough on Recovery Officers as they were on mechanics and electricians.

The spacecraft had rounded the moon and was heading back toward the earth. It was still travelling at over five thousand miles an hour, but the higher it rose from the moon the slower it would go, until the next morning, at the crossover point into the earth's gravity, it would be travelling at less than three thousand miles an hour. At the moment, it was moving so quickly that Haise felt as if he were in a jet plane taking off from a short runway; when he stole a glance out the window, the spacecraft seemed to be rising straight up from the moon. Immediately below, he could see Censorinus, a crater so sharp that it seemed the spacecraft might have just been ejected from it. To the west of Censorinus he could make out Tranquillity Base, where the Apollo 11 astronauts had landed nine months before. He couldn't see the Fra Mauro hills, where he and Lovell had been supposed to land the next day, nor was he ever able to see the crater that Lovell and the other Apollo 8 astronauts had named for him. Brand's voice came in over the crackling radio to report that the booster had hit the moon and made a crater that (on the basis of seismic data) was probably a hundred and twenty feet in diameter, but Haise couldn't see that, either. However, he told Brand he was glad to hear that *something* had worked right on this flight.

Brand was talking to the astronauts less now, for he knew that they were busy. Ten minutes before the burn, Kranz checked with each flight controller in turn to make sure each was ready. At the back of the Control Room, the visitors' gallery was filling with people who wanted to be present; there was even more NASA brass than there had been that morning, for Dr. Thomas O. Paine, who was then the NASA Administrator, and Dr. George M. Low, the Associate Administrator, had flown down from Washington. There was a spectator up in the spacecraft as well, for Swigert was in the LM, looking over Haise's and Lovell's shoulders. Ordinarily, command-module pilots were not present for lunar-module rocket burns. Swigert, who had nothing to do himself, was feeling like a third wheel. Everyone was tense. Brand, who was supposed to say "Mark!" to inform the astronauts when there were exactly forty seconds to go, said "Mark!" by mistake three minutes ahead of time—an understandable mistake, because there were several electronic clocks at the front of the Control Room counting down the time to different events, and it was easy to look at the wrong one. Fortunately, one of the astronauts caught the error.

At the right moment, Brand called out "Mark!" again, and just forty seconds later Lovell, his hand on the throttle, turned on the LM's main rocket. Because—like the free-return maneuver—this was a docked burn, Lovell had to do the throttling manually; even though the computer was what flight controllers called "up and running," it was not programmed for throttling a docked-DPS burn. It would, however, control the guidance and turn the rocket off when the spacecraft had reached the proper acceleration; one of the guidance instruments could measure increases or decreases in speed. Lovell made the burn in three separate stages, so that it could be stopped more easily in case of trouble. First, he throttled the rocket up to ten per cent of its thrust for five seconds, to warm it up. Then he brought it up to forty per cent of its thrust for twenty-one seconds to trim the gimbals. Any problems with the rocket's firing would show up now. In Houston, the Control Officer, Richard Thorson, studied his telemetry. When the CONTROL was sure the burn was going smoothly, Lovell brought the rocket up to full thrust for almost four minutes. The computer turned off the rocket only thirteen-hundredths of a second after the time predicted to reach the right speed, which Deiterich and Spencer thought was surprisingly accurate for a manual docked firing. Now that the astronauts would not be landing in the Indian Ocean, the *Bordelon* was called off, and the special crane was taken off "alert" at Norfolk.

As soon as the PC+2 burn was over, Brand expected Kranz to give him the go-ahead to read off the procedures for powering down the spacecraft for the long coast back to earth. Looseleaf notebooks giving the procedures were spread out on

Brand's console, and he was anxious to get on with them, because he realized how tired the astronauts were, and wanted to get them to bed. CAPCOMS sometimes have to represent the crew's interests on the ground. The go-ahead for the power-down never came, and when Brand turned around to find out why, he saw a number of men standing around Kranz's console. Kranz was in the middle of what he later called "a long and loud debate" between the TELMUS, on one hand, who were anxious to proceed with the power-down immediately, and a group of design engineers, on the other, who were insisting that a good thermal roll be set up right away; the spacecraft's attitude in relation to the sun had been so erratic for about a day, they argued, that some part of the craft was likely to overheat and break down. Having the astronauts roll the spacecraft manually every hour had not proved effective, and the design engineers were hoping to set up an automatic roll, monitored by computer. The TELMUS had powerful allies in members of the astronaut corps, for, like Brand, they all wanted to get the spacecraft powered down as soon as possible, to let the crew have some rest. Donald K. Slayton, who, as Director of Flight Crew Operations, was the unofficial chief astronaut, was telling Kranz that it would take a good two hours to set up a roll, and the astronauts in the spacecraft had had almost no sleep for the better part of two days. The TELMUS were saying that keeping the LM powered up for another two hours would knock their power profile out the window, and that that was nothing compared to what would happen to it if they had to keep the guidance system powered up to monitor the roll. Kranz listened, and then, in the NASA tradition of the top man's making the decision, announced that they would set up the roll anyway—the astronauts, he reasoned, would be better off losing a little more sleep than risking any more breakdowns. They would not, however, keep the guidance equipment powered up all the way back to earth.

Brand put aside the power-down checklist and began reading up the procedures for the roll instead. The LM's guidance system had never been designed to maintain a roll, and doing so was particularly difficult with the two other modules attached. Each time the astronauts got the spacecraft rolling, it developed a wobble, and the wobble worsened until there was no roll left—the spacecraft was like a top at the end of its spin. While the astronauts continued to wrestle with the roll, the FIDO was plotting vectors to get some idea of how accurately the spacecraft had been placed on its new trajectory. David Reed, the Lead FIDO, was worried because there appeared to be an unexplained error in the trajectory. The spacecraft was about one degree from where it ought to be. For a minute he thought he had the answer—he heard Swigert boast that he had been the first Command Module Pilot ever to witness a lunar module rocket burn from inside the LM. When Reed checked with the Control, the two agreed that Swigert's having been out of place was enough to throw the spacecraft off course, as if it were a small boat. Neither Reed nor the Control was satisfied with this explanation, though, because the aberration in the trajectory was a large one, and it was getting worse. *Something* in the spacecraft had to be venting, but Reed couldn't think what.

The astronauts were not told that night about the increasing error in their trajectory. While Lovell kept trying to work the wobble out, Haise began powering down parts of the lunar module. What the TELMUs couldn't win directly, they won by persistence and stealth: they were continually getting Brand to read up small lists of items to turn off. For the two-and-a-half-day coast back to earth, the LM would be almost as inert as the command module. About the only things to be left running would be the radio transmitters and the life-support system.

After an hour and a half, Lovell felt he had set up as good a roll as he could get. Accordingly, when Brand was transmitting some more suggestions from Duke, in the simulator, one of the astronauts in the spacecraft interrupted, "Hey, we've gone a hell of a long time without any sleep," and another voice put in, "I didn't get hardly any sleep last night at all." At about ten-thirty—some twenty-four hours after the accident—Brand told the astronauts that they could begin to power themselves down. This was easier said than done, for the astronauts had built up a good deal of nervous energy. They still hadn't gone to sleep an hour later, and Brand then radioed up instructions from Slayton that they stop fiddling around and get to bed. Swigert, however, who was particularly keyed up, ran through a recapitulation of his impressions of the accident for Brand.

Rounding the moon had given the astronauts' spirits something of a lift, but now that they were on the quarter-million-mile straightaway back to earth, Swigert was growing increasingly apprehensive. As command-module pilot, reëntry would be up to him. He told Brand he was worried about the command module's fitness; in particular, he was concerned about Main Bus B, where the failure in the electrical system had first become apparent. During reëntry, along with Main Bus A, it would be providing power drawn from the three reëntry batteries—assuming the bus itself hadn't been damaged by the accident. "You think Main Bus B is good, don't you?" Swigert asked anxiously.

Brand answered, "That's affirm. We think it is, but we want to check it out anyway. We think you guys are in great shape all the way around. Why don't you quit worrying and go to sleep?"

"Well, I think we just might do that. Or part of us will," a voice said from the spacecraft.

BACK

GENERALLY SPEAKING, SPACE-FLIGHTS ARE LESS SATISfying to the people who follow them on television than the adventures of earth-bound explorers have usually been to those who have read about them in books. For one thing, technology seems to impinge too heavily on the televised space travels; what was once called a spaceship, to give a small example, has become a module. However, technology notwithstanding, the men who ply between the earth and the heavens are not doing anything much different from what was done by the explorers who merely used the heavens to steer by. The Apollo 13 astronauts were now in every bit as understandable and distressing a predicament as any seamen aboard a leaky vessel in danger of foundering. This was readily grasped by the estimated third of the world's population who were following Apollo 13—

probably more than had followed any other spaceflight. There was a sort of worldwide shudder of horror, for if these men died they could do so in a way no men ever had before: they could be the first never to return to the dust of this planet. Newspapers in many countries published cartoons representing the world with a pair of hands; in some cases the hands were reaching out for the astronauts, and in others they were simply clasped in prayer.

Despite similarities with earthbound crises, getting the astronauts back would obviously require a vaster cerebral effort than any conventional rescue operation. After the PC+2 burn, Kranz removed his White Team from the regular cycle of shifts—or, as he put it, he "took it off the line," the way astronauts might set aside a battery that was needed for some special purpose—so that its members could devote themselves entirely to this effort. While the three other teams were handling the relatively routine coast back toward the earth, the White Team would be writing a new reëntry checklist. A checklist for a spaceflight can be as thick as a telephone book and take three months to prepare; now the White Team had to write one in less than three days, and that span of time would shrink by half a day before they actually got down to it, for Kranz wanted his team to get some sleep before they set to work. According to Swigert, the astronauts could never have got back to earth without the White Team's checklist. On that trip back, a number of things would have to be done that had never been done before. In the last hours of the flight, the astronauts would have to leave the lunar module and find some way to power up the command module, the only part of the spacecraft that had a heat shield. A power-up of a dead command module had never even been simulated on the ground, let alone attempted in space, and there were many questions. Would the command module's delicate electronic systems work after more than three days of cold? For that matter, would its three reëntry batteries, which would be supplying its power when it was flying alone, still work after the long chill, or would they be weak and useless, like a car battery on a cold winter morning? Besides these problems, there would be unprecedented difficulties in getting rid of the two other modules just before reëntry.

The necessity for all these new and untried maneuvers meant that the flight controllers couldn't borrow much from the checklists of previous missions—something they normally did—and this inability made them grateful for the extra day the slower PC+2 burn had given them. Doing three months' work in under as many days meant they had to take shortcuts they had never thought possible. Afterward, David Reed, the Lead FIDO, said, "When you take a lot of time, you get the most conservative consensus. But here we shaved off the conservatism to give some fast decisions, and we stuck with them. We found out what we could do when the chips were down. Often, the only simulators we had were our minds,—and damn if they didn't work!"

Early Wednesday morning, Haise was on watch in the lunar module while the others slept in the command module. The temperature in the LM was about fifty-five degrees and falling, for the TELMUS had decided that, during the long coast back to earth, the LM could draw only eleven amperes from its batteries. That way, there would be more electricity to transfer to the command module at the end of the flight.

Jack Lousma, who was back as CAPCOM, did his best to give Haise a little company. Haise told him he could now tell that the spacecraft was definitely moving away from the moon, because, for the first time, he was able to get the entire moon in the field of vision of the LM's monocular, or hand-held telescope. The craft was then over twenty thousand miles away from the moon. Lousma did his best to keep Haise's spirits up—sometimes the CAPCOM had to act as a fourth member of the crew. When the spacecraft wobbled and Haise had to scramble to switch omni antennas, Lousma said soothingly that everyone in Mission Control was one hundred per cent optimistic, and he added, "It looks like we're on the up side of the whole thing now." The up side had its down moments, however. Once, for instance, Haise remarked that he saw a chunk of silvery metal, about four inches square, tumbling around outside the window. Presumably, it was part of the spacecraft's ruptured oxygen tank.

Haise was looking out because, with the guidance platform turned off, the only way he could check on the roll was to time the alternate reappearance at a window of the earth and the moon. From time to time, he would say to Lousma, "O.K., the earth went by there," or "There goes the moon." There was a reticle—a grid of hairlines for measuring positions—printed on the glass, and every time the earth or the moon went by, Haise would look to see if it was in the right spot. It never was; at each pass the moon seemed to be a bit higher and the earth a bit lower, and that meant that the spacecraft was still wobbling. It seemed as if something was venting.

At three Wednesday morning, Lousma heard Lovell's voice on the radio. "Gee whizz," Lousma said, "you got up kind of early, didn't you?" Lovell replied that it was cold upstairs in the command module—that the temperature was somewhere in the forties. This was the first indication that the temperature in the command module, which had dropped irretrievably when Lovell had pulled the blinds, would keep the astronauts from sleeping. Their clothing, designed with the spacecraft's normally conditioned air in mind, was of a flimsy material called Beta cloth, and their sleeping bags were merely thin sheets of fabric designed to keep them from floating around, and were, in fact, perforated for ventilation. It was suggested by the controllers that the astronauts put on their space suits, but they declined, on the ground that the suits would make them too clumsy to handle the spacecraft efficiently. There were some blankets stored near the floor of the command module in an emergency pack to be used in case the astronauts weren't rescued after splashdown and had to head for shore themselves; but the astronauts didn't open the emergency pack because it was so tightly put together that they could never have packed it up again after removing the blankets. If they had opened it, the spacecraft would have been awash in flashlights, tinned foods—even a rubber boat.

At ten-thirty on Wednesday morning, a yellow caution light flashed inside the LM to indicate that the level of carbon dioxide in the spacecraft's atmosphere had built up to the point where something had to be done about it. The Crew Systems Engineers in Houston who had been working on this problem for a couple of days had finally come up with a way of adapting the command-module canisters for use in the LM: There was a hose in the LM meant for sucking air out of space suits, should the astronauts be wearing them, and this could be used to suck cabin air through the canister—provided an airtight connection between hose and canister could be devised. The Crew Systems Engineers had found that they could make the airtight connection out of one of the plastic bags used for storing part of the astronauts' moon-walk garb. The bag could be put over the canister in such a way that the canister's open mesh bottom was outside; then a hole could be cut through the other end of the bag for the hose; and finally, the whole apparatus could be sealed with tape. A major stumbling block had been how to prevent the plastic bag from collapsing with the suction. After experimenting an entire night, the Crew Systems Engineers had solved the problem by holding the bag away from the mesh with one of the astronauts' plastic checklist cards.

The astronauts were very curious about the contraption, for the previous day Vance Brand had given Lovell something of a preview of it by telling him that making it would be like putting together a model airplane, but that when the thing was finished it would look more like an R.F.D. mailbox. During the night, Haise had got together several of the items needed for the project—plastic bags, a roll of tape, a checklist card—but now Haise had gone to bed. Lovell and Swigert were up. Down in the Control Room, the Black Team had just taken over from the Maroon Team, and Joseph Kerwin, who had relieved Lousma at the CAPCOM's console, had got them started on the mailbox. It hadn't been easy to explain over the radio how to build something that no one had ever seen before. "Now, then, we want you to take the tape and cut off two pieces about three feet long, or a good arm's length," Kerwin had begun. Earlier, the Crew Systems Engineers had pondered over the best way to say "thirty inches." "Two or three feet?" "An arm's length?" They had decided to say both. (Wording would prove to be a problem with the entry checklist, too.) Now, when the caution light went on, Lovell turned on the suction. Air from the cabin was drawn through the canister and scrubbed air was vented back again, and in a short time the carbon-dioxide level was back to normal.

The planning for the reëntry checklist got seriously under way at a meeting at about one o'clock Wednesday afternoon, when splashdown—set for around noon Friday—was less than two days away. The meeting had been put off until one in the afternoon in order to give the White Team a good rest following the PC+2 burn the night before. However, because this was the fifteenth of April, several of the flight controllers had spent some of the intervening hours doing their income taxes, which they had left until the last minute because everyone had expected the flight to be a smooth one. The White Team met once again in the unused Staff Support Room on the second floor of the Operations Wing, beneath the big room where the controllers worked and above a computer room on the ground floor. Some of the flight controllers present referred to themselves as the Tiger Team, because the White Team had been augmented by a number of outside engineers, and, in addition, a number of changes had been made in the team assignments, so that the White Team would include most of the Lead controllers. These were the Lead FIDO, David Reed; the Lead GUIDO, Kenneth Russell; the Lead RETRO, Charles Deiterich; the Lead EECOM, John Aaron; the Lead GNC, Buck Willoughby; the Lead TELMU, William Peters; and the Lead CONTROL, Harold Loden.

What the Tiger Team was planning was the

last six hours or so of the flight—the crucial period when the crippled spacecraft had to be made ready for the plunge through the atmosphere. Kranz, the Flight Director, standing at the front of the room, briefly outlined what he thought were the most important points: The controllers would have to be completely accurate, because the spacecraft would be less flexible and therefore less tolerant of error than ever before. They would have to develop procedures in such a way as to let the LM do most of the work. There might be many ways to do the job but only one that was best, and that was the way that would be easiest on the consumables and easiest on the astronauts. The talk was over quickly, for Kranz's words tended to be terse, like those of the captain of a ship caught in a squall.

Kranz felt that the key to the problem was to conserve the command module's reëntry batteries as long as possible by having the lunar module supply all the electricity for the initial powering up and warming up of the command module; only in that way was there a chance that the reëntry batteries would last through the critical period between the jettisoning of the LM and the splashdown—the period when the command module would be flying alone. Normally, the three reëntry batteries were expected to provide power for about forty-five minutes, but now it was a question of doing so for several hours. Moreover, there were only two and a half reëntry batteries this time, because half of one had been used up after the accident, and, all told, they could provide only ninety-six ampere-hours of electricity. (One amp-hour of electricity, on the spacecraft's twenty-eight volt current, would keep a forty-watt bulb burning for one hour.) Obviously, then, the more the LM could do for the command module ahead of time, the better. Since everything depended on the transfer of power from the LM to the command module, and since such a transfer had never been tried before, as soon as he had adjourned the meeting Kranz beckoned to three engineers. One, from the Grumman Aerospace Corporation, was a chief designer of the LM; another, from the North American Rockwell Corporation, was a chief designer of the command module; and the third was the Lead EECOM, John Aaron, who would be in charge of the power-up. Kranz wanted to get these men together because they had most of the necessary information in their heads: Aaron knew what had to be powered up, the North American man knew how much power it would take, and the Grumman man knew how to deliver the power from the LM. Later, some of the flight controllers said Apollo 13 was "an EECOM's mission" equally as much as a RETRO's. Aaron, a tall, ruddy-faced twenty-seven-year-old graduate of Southwestern State College, was a veteran of many emergencies—he had been the man who had made the decision to go ahead with the Apollo 12 mission after the spacecraft had been hit by lightning. He said later that the members of the Tiger Team were all perfectly well aware of how little time they had before the astronauts smacked into the atmosphere but that nobody in the room seemed to have a clear idea of where to begin. People seemed to be wandering aimlessly about. "Suddenly I realized that it all had to begin with *me,*" Aaron recalled. Since he knew he couldn't formulate a coherent plan with so large a group, he collected four or five other flight controllers and took them, along with the men from Grumman and North American, over to a quiet corner. Kranz said later that there were only a few things he would do differently if a similar crisis arose again, but one would be to pack off small groups of engineers to other rooms instead of having them all milling around in the same place.

Each of the engineers with Aaron had a chart of the command module's dashboards—one of dozens of charts that had been passed out so that the flight controllers could jot down their ideas for different arrangements of the switches, much as a choreographer might block out the steps for an intricate dance. The power-up was certainly intricate, for a switch might have to be left on for one action and then turned off before the next could begin. At nine that morning, Houston had read up to Swigert the basic arrangement of the switches at liftoff, and it was this arrangement that was printed on the charts. The engineers called it the Square One arrangement, and in figuring out different checklists they always returned to the Square One arrangement; otherwise they and the astronauts would have become hopelessly muddled. With the charts, Aaron and his group began to work on what Aaron called the "strawman timeline"—his term for a rough outline of what had to be done ("A strawman is something you can throw rocks at," he explained), and how best to use the available power, in particular just before splashdown. The strawman timeline included estimates of when different parts of the spacecraft should be powered up, when the service module would be jettisoned and when the lunar module would be, and how much power should be allocated to each operation. When Aaron had finished the initial calculations, he concluded that the command module would have a margin in its batteries after splashdown of just sixteen amp-hours—none too much, for the Recovery Officers were insisting that there must be enough power to inflate the balloons that could right the spacecraft if it landed upside down, and to power a radio beacon signalling its location when it landed. Aaron's strawman timeline was very rough, and would undergo many changes, but at least it provided a framework that other flight controllers could fit their more detailed checklists into. "Things always overwhelm you until you have a plan," Aaron said afterward. Even with a plan, there were to be moments when Aaron and the others felt overwhelmed.

At about ten Wednesday morning, Haise, who had been napping in the command module, came into the LM complaining of the cold upstairs. One source of his discomfort, it later turned out, may have been the fact that he was coming down with a kidney infection. It was not the best of mornings.

Haise looked mistrustfully at the lithium-hydroxide mailbox that had been built while he was asleep. It added to the clutter in the spacecraft, which was beginning to fill up with polyethylene bags containing the astronauts' wastes; normally, these would have been vented through a special valve in the command module, but this was now probably frozen, and also such venting could disturb the craft's fragile trajectory—Reed used to tell astronauts there wasn't much he didn't know about them. Yet even though apparently nothing was being vented overboard except the negligible amount of water used to cool the electronic instruments, the craft's motion continued to be erratic. Haise noticed that the wobble in the thermal roll was worse than it had been the night before when he went to bed. At each revolution, the earth appeared so high in the window that he almost had to get down on the floor to see it, and, conversely, the moon was so low that he had to bob up toward the ceiling to catch a glimpse of it.

In the third-floor Control Room, the FIDO was getting uneasy because the spacecraft's trajectory had been shifting from its predicted course ever since the PC+2 burn; specifically, the trajectory was shallowing out, and this meant that the angle at which the spacecraft would hit the earth's atmosphere was becoming more and more acute. If it got too shallow, the spacecraft would miss the narrow downward corridor it had to hit for reëntry and would skip back out into space, in an orbit so elongated that it would return to the vicinity of the earth perhaps in two weeks, perhaps never. Since no one could account for the shallowing, no one could predict whether it would get better or worse. But, whatever was causing it, something had to be done about it before the astronauts missed the corridor—and the earth—altogether. Accordingly, the FIDO and the RETRO decided on a midcourse correction, to be made at about nine-forty-five that night.

On the second floor, the Tiger Team engineers, on their own initiative, now began breaking up into small groups and moving into adjacent rooms. There were so many such groups that Reed, the Lead FIDO, had difficulty finding a suitable nook in which to hold a meeting of the Trench. Then, when he finally did find one, Deiterich, the Lead RETRO, whom he was supposed to be meeting with, had dashed off to another meeting. Kranz kept moving from one group to another, for he felt that his main job was to make sure all the engineers were working in the same direction, and that the right matters were taken up by the right people. Another engineer listened in on conversations and jotted down ideas to be integrated later. There were some forty men involved, each with his own particular area of responsibility, so uniformity wasn't always easy to achieve. Kranz later attributed their success to the intensive drilling they had gone through for months ahead of the flight. Although the flight controllers had never handled either a simulation or a theoretical approximation of anything like the present situation, they had been obliged to solve a great many problems in the course of dozens of simulations, and they had met and divided into groups to do so. During those months of simulations, channels of communication had developed between them, so within seemingly random discussions there was a sort of order, the way there is between the cells in a computer or those of the brain.

On Wednesday afternoon, the flight controllers began making preparations for the midcourse correction to counteract the shallowing. It would be called Midcourse Correction 5, or MCC 5—a designation referring not to any consecutive number of rocket burns but, rather, to one of several points along the trajectory where a midcourse correction was normally made if it was needed. It wasn't an exact point, and that was fortunate, because the CONTROL had noticed a problem that might require a postponement of the MCC 5 burn. Inside a tank of supercooled helium (used for forcing propellants out of the fuel tanks and into the rocket engine) in the lunar module, pressure was building up to the point where the tank would vent before long. Because this could knock the spacecraft off course again, the CONTROL told the FIDO and the RETRO that it might be best to delay the burn until the venting was over. The trouble was that the CONTROL did not know for sure when the venting would take place. A small disc inside a safety valve would rupture when the pressure reached approximately nine hundred pounds per square inch, but no one could predict the precise instant when this would happen; the best guess seemed to be that it would vent at about ten-forty-five—not long after the time for which the midcourse correction was scheduled. In theory, the venting should not disturb the spacecraft, because the helium would blow out through two apertures facing in opposite directions—an arrangement called a "non-propulsive vent." The CONTROL told the RETRO that he didn't have much confidence in the vent's non-propulsive qualities, so the RETRO agreed to delay the burn for an hour —from nine-forty-five until ten-forty-five, so that the LM would be powered up, and the astronauts would have control—just in case.

At two-thirty in the afternoon, not long after this decision had been made, Haise reported hearing a loud thump in the spacecraft, and he told the CAPCOM, Vance Brand, that he saw through the window a "shower of snowflakes," which looked as if they had come from the vicinity of the helium tank. The CONTROL, who was almost as startled as Haise, had the telemetry turned on. He reported that the helium still appeared to be in the tank. Haise was not particularly reassured, and neither was the CONTROL, for if the helium hadn't made the thump and the snowstorm something else had. Haise looked up and saw that the moon had moved to the docking window overhead; this meant that whatever had thumped had caused the spacecraft to wobble some more. The occurrence was uncomfortably reminiscent of the original accident. A couple of hours later, Lovell reported that there had been a master alarm indicating trouble in one of the LM's batteries. Many of the flight controllers were strongly reminded of the caution light indicating trouble in Main Bus B that had accompanied the initial explosion two days before, and they became extremely quiet. Much later, the flight controllers determined that the thump had been made by a malfunction in one of several batteries in the LM. As if history were repeating itself, a supposedly automatic switch had failed and remained on; there had been a short circuit inside the battery; and an electric arc had opened the battery vent valve and sent a shower of liquids into space. This time, however, the damage was much less serious. The TELMU asked Lovell to take the battery off the line for the time being, but it would continue to work after the short circuit even with its vent open.

Then everyone sat back and waited for the helium tank to blow.

David Reed, the Lead FIDO, and Charles Deiterich, the Lead RETRO, got together in the Flight Dynamics Support Room, on the second floor, and began working out the steps leading to splashdown. Deiterich outlined the plan he had thought up the morning after the accident for jettisoning the service module: The LM would push the entire spacecraft with the service module in front; then the astronauts would detach the service module and reverse directions. Deiterich said it would be as easy as stepping out the door to throw away some garbage and then stepping back inside. With the service module gone, the next major item would be bringing up the command module's guidance computer and aligning its platform. Kenneth Russell, the Lead GUIDO, who had recently taken this matter up with the Lead GNC at the behest of Kranz, reported that they had been afraid the astronauts once again would be unable to see clearly enough through the windows to distinguish the guide stars. He proposed that in the event they couldn't make a star check, they base the alignment on the sun and the moon instead of on two stars. Deiterich and Reed didn't care much for the idea, since they felt that two imprecise checks weren't much better than one, but they said they could get by with it if they had to.

The next item would be jettisoning the lunar module—normally an easy job, because it was done in lunar orbit, where there was plenty of time, and where the service module could push itself and the command module away from the LM. Now the service module would be gone, and there would be no second chance, because the spacecraft would be hurtling toward the earth. The morning after the accident, Deiterich had also blocked out a way to do the LM jettison, and he recapitulated it now: When the astronauts shut the hatch of the lunar module and also that of the command module, air at cabin pressure would be trapped inside the tunnel between the two; then, when the docking mechanism was released, the pressure would blow the two craft apart, in a sort of cosmic sneeze. Reed and Russell liked the idea. So did Peters, the Lead

TELMU, and Loden, the Lead CONTROL—the two officers who were responsible for conserving the LM's power and propellants—when Deiterich told them about it a little later, since the plan required absolutely no LM maneuver whatever. The LM Systems Engineers were having their own meeting next door—they were extremely interested in knowing what the Trench Engineers had in mind, so that they could budget the electricity, and once they waylaid Deiterich for so long that the Trench meeting came to a halt. Then, by the time he got back, some CM Systems Engineers had made off with Russell—"People were always dragging you off," he recalled later. "Everyone had a question for everyone else, and before you finished one meeting, you found yourself in another."

When the Trench group was back together again, Russell brought up another navigational worry. Just before reëntry, the command module—by itself now—would be coming around the dark side of the earth, and that meant that the astronauts wouldn't be able to check their trajectory with reference to the earth's horizon, as they normally did. Deiterich had a thought. Instead of doing a horizon check, he said, perhaps the astronauts could do a "moonset check"—a term he had invented—which would involve noting the moment the moon disappeared over the horizon, to see if it vanished at the predicted instant. Russell said he would look into it.

Upstairs in the third-floor Control Room, the console immediately to the left of the CAPCOM's was occupied by the Chief Flight Surgeon, Dr. Willard R. Hawkins. He and the other Flight Surgeons had not been very busy since the accident, because the biomedical sensors, which normally sent back information on the astronauts' pulses, their breathing, and their temperatures, had been among the first parts of the telemetry to be turned off after the initial accident. Thereafter, the Flight Surgeons had had to rely for medical data on whatever they could pick up from the astronauts' conversation. It had occurred to Dr. Hawkins that the astronauts had probably not been drinking enough water—once a common problem among people in ships' lifeboats—and from time to time he asked the CAPCOM to urge them to take a drink, but he could not be sure whether they actually did. One difference between spacecraft and lifeboats at sea is that astronauts in space don't feel particularly thirsty, even when they badly need water. Some experts say that in the earth's gravity blood tends to pool in the lower legs and feet, but in zero gravity this doesn't happen, so there is more blood around the chest and heart, which the body has to diminish; it does this partly by increasing elimination and partly by decreasing thirst and, hence, the intake of water. Consequently, astronauts can get seriously dehydrated without knowing it. Because the shortage of water aboard the LM was critical—water was essential for cooling the electronic systems—the astronauts had been rationing their own intake severely and saying nothing about it to the ground, a move that complicated Dr. Hawkins' analysis of their medical problems. However, the doctor had a trick up his sleeve, or thought he had: There was a big tank of water in the service module, and he suggested that the astronauts might drink from that. The difficulty was that to get the water out, the astronauts would have to use some of the sparse supply of entry oxygen from the surge tanks, and they weren't thirsty enough to do that. One of the Crew Systems Engineers also remembered that the space suits, which the astronauts weren't wearing, were threaded with tiny capillaries full of water, and all an astronaut would have to do would be to snip off a space-suit toe and drink from it as if it were a wineskin. The astronauts weren't thirsty enough to try that, either.

As the flight controllers developed their plans, they were continually intercepting some of the several astronauts who remained on hand to help, showing them charts of the dashboards on which they had figured out some new arrangement of switches, and asking their opinion about whether a projected maneuver would actually work. Whenever the flight controllers were far enough along with a plan so that they wanted to have it tried out, they sent one of the dashboard charts, with appropriate notations, down to the astronauts on the ground floor who were running the simulators. Charles Duke, working with the LM simulator, was running through Deiterich's proposal for jettisoning the service module. Deiterich had stipulated that the spacecraft be brought to a specific attitude first, for otherwise the service module might collide with the rest of the spacecraft farther on; but Duke, wrestling with this change of attitude in the simulator, could never be sure it accurately duplicated the movements of the LM maneuvering the other two modules. Duke moved on to Deiterich's scheme for jettisoning the lunar module by letting it sort of fizz away, propelled by the pressure in the tunnel. It was complicated by the fact that a lunar module and a command module had never flown alone before. As the LM's attitude indicators would be off, he had to maneuver to the right attitude for jettison by centering one after another of a series of guide stars in the window, and he found that the series of stars kept bringing the LM's inertial platform close to gimbal lock. Duke requested a new set of stars, and he complained that acquiring the new attitude would take too much time. Deiterich, however, insisted that it was important, because the LM carried a metal cask full of radioactive fuel, which had to be aimed so that it landed in deep water. The radioactive fuel had been intended for powering scientific instruments left on the moon, and its cask was the only part of the LM that was strong enough to remain intact during the heat of reëntry; it had been designed this way because during a previous mission such a cask, aboard an unmanned satellite, had scattered its radioactive contents through the upper atmosphere, and four years later these could still be detected around the world. Earlier in the Apollo 13 flight, when it had seemed possible that the spacecraft would come down in the Indian Ocean rather than in the Pacific, the Atomic Energy Commission had sent anxious word that the fuel cask would land uncomfortably close to a populated area of Madagascar. Now Deiterich had assured a representative of the A.E.C. who was present that the controllers would see to it that the cask landed in deep water a couple of hundred miles off the coast of New Zealand.

The White Team's work on the reëntry checklist continued without letup into Wednesday night. One of the busiest flight controllers was John Aaron, who said later that the time he spent on the checklist was an unbroken blur of discussions, simulations, and revisions—a single continuous process. Whenever one of the other engineers wanted to turn an instrument on in the spacecraft he had to find out whether Aaron would give him the power. Accordingly, Gary Coen, one of the GNCS, who were responsible for maintaining the navigation and propulsion systems that would keep the command module in the narrow reëntry corridor, sought out Aaron to see how much power he could pry loose. The command module had three separate navigation systems, and clearly they couldn't all be powered up. Coen wanted to use its Primary Guidance and Navigation System, or PGNS, which included the guidance platform and computer and was the most reliable, but Aaron wouldn't hear of it; the PGNS used too much power. Coen then asked about the secondary system, which included a set of six small gyroscopes that did the same job as a guidance platform but not as reliably. The secondary system did use less power, but Aaron was not sure he had enough power even for it. The third alternative was a simple meter that indicated gravity forces, or Gs. The astronauts would know they were within the reëntry corridor if they were receiving the right number of Gs, which would increase at a known rate all the way down. If the command module began receiving too many or too few Gs, they could bring it back on course by manually adjusting its tilt so that the command module's flat bottom bit more or less deeply into the atmosphere. When the command module was receiving Gs at the same rate once more, the astronauts would know they were back in the corridor again. The simplicity of this expedient was a link with the days when explorers had no computers or inertial platforms to guide them, but, simplicity or no, none of the flight controllers were happy at the thought that they might actually have to use this method, which they called "riding down the Gs." Coen didn't want to try it unless he had to. About the only concession he could get from Aaron was that, in addition to the gravity meter, *half* of the secondary system—three of the six little gyros—could be powered up early enough to be warm at the time of reëntry. That was not much of a guarantee, but Coen let the matter go for the time being. He had to hit Aaron for electricity for another purpose, to turn on some heaters which would warm the command module's thruster jets in case propellant had frozen inside their nozzles; if the ice wasn't melted, the thrusters might have difficulty firing while the astronauts were trying to keep the command module within the corridor, regardless of which navigation system was used. Aaron said no. Coen then offered a suggestion that would require less power, that the thrusters be de-iced by test firing them. They consulted Reed, Russell, and Deiterich, who objected at once that the test firing could knock the spacecraft off its trajectory. At times the spacecraft seemed so hopelessly complex that no flight controller could propose an action without several others vetoing it. Even Loden, the Lead Control, who overheard the discussion, complained that it might bring the LM's guidance platform close to gimbal lock. Almost anything, however, was better than a plan Loden and some of the LM systems engineers were hatching, which was to be used in the event that the command module couldn't be powered up at all, so there wouldn't be even a gravity meter, let alone thruster jets. These engineers had figured that if worse came to worst, the LM could place the command module at the right attitude for reëntry and then, before it was itself jettisoned, set the command module spinning at about four revolutions a minute—fast enough to stabilize it as it plunged through the atmosphere.

A little before ten Wednesday evening, the astronauts began getting ready for MCC 5, the midcourse correction to counteract the shallowing; it would be a small burn, designed to shave seven feet per second off the spacecraft's speed. The spacecraft had crossed over from the moon's gravity into the earth's gravity that morning, and had since been slowly picking up speed as it began the long drop into the Pacific. It was still travelling at less than three thousand miles an hour, but in a day and a half, when it hit the atmosphere, it would be going more than eight times as fast as that.

Everyone was still waiting for the helium tank to vent, which it still hadn't done, but Deiterich said he couldn't delay the burn any longer. Now Lovell found the earth in the telescope. At that time, the earth's terminator caused it to have horns, like a crescent moon, and Lovell had been told to align one of the cross hairs so that it was just grazing the horns. "I hope the guys in the back room who thought this up knew what they were doing," he grumbled. Reed, who had been the one to suggest this alignment, wasn't worried. It was necessary to orient the spacecraft in two directions, and lining up the earth's horns on one of the cross hairs would do the trick, because the shadow was always at right angles to the sun; without this trick Deiterich would never have let the TELMUs turn off the LM's guidance system on the trip back. The method had been thought up as an emergency measure at the time of Apollo 8, the mission in which Lovell and his companions had made the first manned lunar-orbital flight, and when the procedures were read up in the afternoon of April 15 he couldn't believe what he was hearing, because even when he had first heard them during Apollo 8 he hadn't dreamed he would ever have to use anything as risky as that. Meanwhile, Deiterich was battling with the weathermen, who kept predicting that the tropical storm Helen was approaching the landing site, and who wanted Deiterich to shift the target several degrees to the west. Deiterich said later that he "hounded those guys all over again." He refused to budge the landing site. The maneuver PAD for the burn was already copied down in the spacecraft, and, besides, he felt he could fly over any storm the weathermen could produce simply by tilting the spacecraft's attitude at reëntry so that it skipped more in the atmosphere.

The CAPCOM was as nervous about the burn as the astronauts, for when Lousma, who was now on duty, called out that there were three minutes to go, Lovell corrected him, pointing out that there were just *two* minutes—Brand had made the same mistake before the PC+2 burn. While they waited, the CAPCOM made sure that Swigert was sitting in exactly the same spot in the lunar module that he had occupied during the burn the day before—this time, the FIDO and RETRO had planned the burn with Swigert's whereabouts in mind. Because the burn was such a small one, the big DPS rocket in the lunar module would be run at only ten per cent of its capacity, and the steering would be done with the thruster rockets instead of by aiming the DPS. With all the automatic equipment powered down, everything was done manually. All three astronauts were needed in this process, and it was as though several people were driving a car at the same time—one working the brake, one the accelerator, and one steering. Swigert kept track of the time for Lovell, and at Swigert's word Lovell pushed the button that started the rocket. Haise kept the telescope cross hair just touching the horns of the earth, and made sure it stayed there by adjusting the spacecraft's pitch, roll, and yaw with the hand lever that fired the thrusters. Since the spacecraft was perpendicular to its trajectory, the burn made it skid like a speedboat on a turn. The burn was completed at 10:31 P.M.

At about two o'clock Thursday morning, the helium tank finally blew out. The CONTROL noticed it almost as soon as Lovell did, and had Lousma ask him if he could see anything. Lovell reported lots of what he called "sparklies" going by outside the window. The spacecraft was wobbling about like a toy balloon jetting through the air. Lovell asked if that was really what they called a non-propulsive vent, and Lousma, after hearing what the effects had been, said he'd hate to see a propulsive one. When it was over, Lovell had to set up the passive thermal-control roll again. As he tried to get the earth and the moon following each other in the window, he and Lousma—both quite bleary-eyed—had a hard time deciding whether it was the earth or the moon that Lovell saw floating by.

"The moon went by the window at six degrees," Lovell reported to Lousma.

"O.K. Earth at plus six. Thank you," Lousma repeated.

"That's the moon. The *moon!*" Lovell corrected him.

"O.K., the moon. Thank you," Lousma acknowledged.

A little while later, Lovell himself became confused about which it was that he was seeing. "The moon passed by at a minus-eight degrees," he told Lousma, and then quickly corrected himself. "No, that's the earth. The earth passed by at minus-eight degrees."

"O.K., the moon went by at minus-eight degrees," Lousma acknowledged, and Lovell was too tired to correct him.

When the CAPCOM asked about the other crew members, Lovell reported that Haise was lying in the tunnel with his head on the ascent-engine cover and Swigert was resting on the floor, tied down with a restraint harness. It was now three in the morning.

A few hours earlier, Aaron had presented his strawman timeline to a meeting of the entire Tiger Team. He jotted down the main points on a blackboard that had been set up at the front of the Support Room. Almost immediately, Commander John Young, the astronaut who was in command of the backup crew for Apollo 13 (and who, along with Duke, landed on the moon two years later, in April, 1972, during the Apollo 16 mission), raised his hand to object that Aaron had set the jettison of the service module too close to the jettison of the lunar module. The period between the two would be the busiest for the astronauts, he said, and they would need more time. Kranz agreed, and Aaron compromised by setting the service-module jettison an hour earlier. The meeting turned into an intricate bargaining session between the other flight controllers, who needed power, and Aaron, who didn't want to give it to them. Sometimes Aaron won a clear-cut victory, but usually the results were mixed. Aaron wanted to delay starting up the command module's instrument-cooling system until just an hour before reëntry, but Willoughby wanted to turn it on two and a half hours ahead of time. Aaron compromised at an hour and a half. He then had to find two amp-hours to keep the command module's air-purifying system running for thirty minutes, because Dr. Hawkins pointed out that there might be pockets of carbon dioxide that the air-purifiers had missed. What with losing a few amp-hours here and gaining a few amp-hours there, Aaron was never exactly sure where he stood, but when the meeting ended he figured that he still had the sixteen-amp-hour margin that the Recovery Officers needed after splashdown.

After the flight controllers had revised the strawman timeline, they had to fill it out, smoothing the details, making sure that actions didn't conflict, and blending them into a unified whole. The checklist would go through several revisions before Kranz was satisfied. In the simulator, Lieutenant Commander Thomas K. Mattingly—whose susceptibility to measles had led to his replacement on the flight by Swigert, and who two years later would be the command-module pilot for Apollo 16—tried out the timeline with Commander Young. They were on the lookout for technical errors; to everyone's surprise there were none. Occasionally they crossed out an item they found hard to understand and replaced it with new wording.

Aaron had a bad scare, for Young and Mattingly, when they were finished with their first run-through, reported that instead of a margin of sixteen amp-hours there was a deficit of ten. Aaron began looking over the timeline to see what else could be cut, but as he was doing so Peters, the Lead TELMU, sought him out to tell him he could have a little more power from the LM. As the astronauts neared the earth, more and more possibilities for trouble were ruled out; for example, there was no longer any possible chance that the flight would last an extra day—an eventuality the controllers had been saving consumables against. The TELMUS, however, would release only a little power to Aaron, for something could still go wrong—a point that Aaron was in no position to argue, since he liked to use it himself with flight controllers who wanted him to cut into his reserve electricity supply. However, Peters did release enough power to Aaron so that he could recharge the reëntry battery that had been used the night of the accident and was twenty amp-hours short of capacity. He suddenly found himself with a margin of ten amp-hours.

The astronauts became increasingly uncomfortable as the flight continued. The temperature within the command module had reached thirty-eight degrees. The astronauts stopped referring to the command module as "the upstairs bedroom" and began calling it "the refrigerator." Some hot dogs they found there seemed to be frozen, they said. Lovell reported that any metal he touched conveyed the cold of space; it seemed to draw heat from his body and dispel it to the stars. Haise, with his kidney infection, was chilled to the bone; once, he shivered for four hours straight. Swigert's feet were still cold because he had spilled water in his shoes two days earlier; the other astronauts, who were equipped for walking on the moon, had pulled on their lunar overshoes, but Swigert had none. Once again Lovell refused a suggestion from the ground that the men put on their space suits; not only would the suits be impossibly clumsy but because of the shortage of power they couldn't be ventilated; the astronauts would perspire and might catch pneumonia. The walls and windows were dripping with water that had condensed from the cabin air; it was like an icy rain.

The cold in the LM prevented the men from sleeping, and lack of sleep made them feel even colder. Sleeping in the LM was further prevented by noise from the craft's instrument-cooling system, which rattled and hissed like the pipes of a locomotive as the water slowly steamed into space. Moreover, a napping astronaut was bound to wake up whenever one of his mates was talking to the

CAPCOM; without an amplifier, an astronaut had to shout to make the CAPCOM hear him. The Flight Surgeons worried, because sleeplessness could reduce the astronauts' efficiency. During the three and a half days between the accident and the landing, Lovell had eleven hours' sleep and Swigert had twelve—an average of about three hours a day, and none of it was what either astronaut could describe as "good" sleep.

Sometimes Dr. A. Duane Catterson, the Deputy Director of Medical Research and Operations at the Manned Spacecraft Center, would join the Flight Surgeon at his console, and the two would listen in silence to the astronauts' voices. The doctors worried as much about the astronauts' systems as the flight controllers did about the spacecraft's. The cold and the exhaustion wouldn't seriously harm the astronauts, but lack of water would. A water tank in the command module, which Dr. Hawkins had thought the astronauts could tap, had now frozen solid; the spacecraft was carrying an internal iceberg. In the four days between accident and splashdown, each of the astronauts drank about a pint and a half of water; they were allowing themselves only six ounces a day—less than a fifth of the normal daily minimum requirement of thirty-two ounces. Dehydration alters the body chemistry with respect to blood, enzymes, and steroids in a way that makes it more difficult to cope with emergencies, but since the astronauts were not particularly thirsty, they would have no warning of this situation. Dr. Catterson worried specifically because water is essential for dissolving certain electrolytes in the body that are important for the transmission of impulses along nerves; without liquid, thought and movement deteriorate. Dr. Hawkins, for his part, worried because the body excretes more potassium when it is dehydrated, and potassium is a vital electrolyte in brain cells. Both doctors feared that the astronauts would start making errors, and, as it turned out, their fears were justified.

Dehydration was a problem in another way as well, for electrolytes are part of the body's defense against disease, and both Dr. Catterson and Dr. Hawkins worried because lack of water could make the astronauts more susceptible to infection. The likeliest place for such an infection, they knew, was the kidneys, which try to retain fluid and consequently never purge themselves. The doctors were unaware that Haise was already developing a kidney infection. Afterward, Dr. Catterson said that if the flight had lasted much longer Lovell and Swigert would almost certainly have developed similar infections, or infections of other sorts, and the astronauts would have been increasingly unable to cope. In their exhausted condition, their nervous energy would have deserted them at some point, and, in spite of themselves, they would just have drifted off to sleep.

As for what the astronauts were actually feeling around that time, it was as much as anything an incessant worry about their reëntry to earth. The flight controllers were worried, too. Fears crowded in from all directions. Around three o'clock on Thursday morning, a dozen members of the White (or Tiger) Team gathered at the CAPCOM's console in the third-floor Control Room while the CAPCOM asked the astronauts some questions that the White Team had prepared about the reëntry, which was still projected for noon Friday. In the spacecraft, Lovell was the only one up. The White Team was not absolutely certain that the seal of the command module's hatch was airtight, and the flight controllers had to find out whether Lovell wanted the crew to wear space suits during the reëntry. If the hatch wasn't tight, the astronauts would lose their oxygen. (A failure of this sort was to kill three Russian cosmonauts in 1971.) They would run a test on the hatch's integrity just before reëntry—a standard procedure—but at that point there wouldn't be enough time for them to get into their space suits if the hatch was found to be leaking. The White Team suggested to Lovell that the astronauts put on their space suits at the very beginning of the reëntry procedures, some six hours before splashdown. Lovell was worried that the ground knew something about the hatch that he didn't know, but Lousma assured him it didn't. Lovell pondered the matter for some time, and in the end he decided against the space suits, for the same reasons as before—their bulk and stiffness.

There was another concern. At about four-thirty in the morning, when Lousma was reading up to Swigert the procedures for recharging the entry battery in the command module—the recharging would take fifteen hours and require transferring electricity through an umbilical from the LM to the command module—Lovell came on the radio to say he was worried because the operation required reversing electrical currents from their normal paths. Usually it was the *command* module that fed power to the *lunar* module, and Lovell feared that the reversal would cause short circuits. Yet the entire entry procedures depended on transferring power from the LM to the command module in order to power up its equipment five or six hours ahead of time. The EECOMs and the TELMUs thought they had figured out a way of getting the circuits to do what they wanted, though it had never been tried before. Lousma tried to be encouraging.

As Thursday dawned in Houston, the astronauts in the spacecraft were trying to get some rest. At about eight o'clock, Kerwin, who had replaced Lousma, accidentally pressed a switch in front of him, called the "king switch," that set off a loud beeping in the spacecraft—a signal that the astronauts were wanted on the radio. Swigert answered, and Kerwin apologized for bothering him. In the desultory conversation that followed, Kerwin told the astronauts about another worrisome matter: the FIDO was tossing around the idea of doing another midcourse correction, five hours before reëntry. Unaccountably, the trajectory was shallowing out again.

HOME

JUST BEFORE NOON, LOVELL RADIOED DOWN TO KERWIN that, with about twenty-four hours to go before splashdown, it would be nice to have the reëntry checklists up in the spacecraft, so that he and Swigert and Haise would have a chance to look at them. The day before, Kerwin had told the astronauts that the lists would be ready "by Saturday or Sunday at the latest." Kerwin was now able to tell Lovell that the checklists were in good shape ("They exist," he said) but that there was still some work to be done on them. By noon Thursday, one final job remained: the checklist for the lunar module and the checklist for the command module had to be tested out together in a computer called the CMS/LMS Integrator, which, as its name implies, would insure that the instructions for the two modules dovetailed—that there were no conflicts between them. This final simulation, which was being run by Mattingly, would take at least four hours more.

The astronauts had plenty to keep them busy while they were waiting. They had to set their cameras to photograph the service module as it was jettisoned the next day. They had to move equipment that was to be brought back to earth (such as the film they had taken behind the moon) from the lunar module into the command module. Then, they had to transfer quite a lot of unneeded equipment from the LM into the command module, because Deiterich was afraid that without the hundred-odd pounds of moon rocks they had expected to have with them the command module would be unbalanced on its plunge through the atmosphere. Lovell, who didn't want a listing spacecraft, either, kept asking the ground to think up more heavy items for them to take into the command module. As fast as Lovell moved such items into the command module, Haise was moving other items out. Over the past four days, the astronauts had been storing their garbage—discarded lithium-hydroxide canisters, bags of wastes—in the command module, where it wouldn't be underfoot, but now that the command module was being made shipshape, the garbage had to be moved back into the LM. "Boy, you wouldn't believe this LM right now! There's nothing but bags from floor to ceiling," Haise said when the job was done. He snapped some pictures for the record.

The work warmed the astronauts a little, but

when they stopped they grew cold again. The temperature in the LM had fallen to forty-five degrees—only ten degrees higher than that now prevailing in the command module. The CAPCOM said it sounded like a cold winter's day and asked if it was snowing yet.

At about six-thirty Thursday evening, or less than eighteen hours before splashdown, Vance Brand, who was back on duty as CAPCOM, said that the checklist was about ready to be read up. The members of the Tiger Team were assembling in the third-floor Control Room to listen in; their work on the checklist wouldn't be complete until it had been delivered to the astronauts. Brand told Lovell that the readup would start with the part of the timeline that dealt with the command module, so Lovell put Swigert, the command-module pilot, on the radio. "He'll need a lot of paper," Brand said. Earlier, Haise had mentioned an item nobody had thought to provide—"a big book with a lot of just plain old blank pages in it." Swigert had been scrounging blank pages from other checklists in the spacecraft.

When Swigert had pen and paper ready, Brand told him to wait a minute; Aaron, the Lead EECOM, wanted a copy of the checklist, so that he could follow it as it was read up. Swigert was irritated by the delay; he knew that there had never been a checklist quite like this one, and he wanted to get on with the job of copying it. To pass the time, Brand recited the titles of the different checklists spread out on the console in front of him: one for the service-module jettison, one for the lunar-module jettison, and one for aligning the command module's guidance platform. As Brand at last seemed about to start the readup, he broke off again, explaining that some more of the engineers who had worked on the checklist were coming into the Control Room. Deiterich, Reed, and Russell arrived from the ground floor, where they had been practicing exactly what they would be doing in the Trench as they guided the astronauts in. To the astronauts, the wait seemed interminable. The checklist was their only passport back to earth, and the earth was looking closer, and more solid, every minute. In reality, they were still about a hundred thousand miles away—not much more than half-way back—but they were travelling at four thousand miles an hour now, and the increasing pull of the earth's gravity would get them back in about seventeen hours.

While Brand waited for the Tiger Team engineers to find seats, he "voiced up"—in the NASA phrase—some housekeeping details. Several more minutes passed, during which Brand asked Swigert for some read-outs from the battery that was charging, but before Swigert could comply Lovell burst in to say, "We can't just wait around here to read the procedures all the time up to the burn! We've got to get them up here, look at them, and then we've got to get to sleep!" Brand said everybody should be ready to start the reading within five minutes; there weren't enough copies of the checklist for all the engineers that had just come into the Control Room, and someone had gone to get more.

Kranz was aware that the astronauts were tired and impatient, but he went along with the delay. "I knew communications were getting terse," he said later. "Everyone gets to that point. It's nothing personal—just frustration. When you're passing data to the crew, there's no substitute for accuracy. It's difficult to know what the proper trade-off is, but the key thing is to be right."

At last a door opened and several more engineers entered the Control Room, bearing stacks of multilith copies of the checklist, and these were passed out. The last to enter was Lieutenant Commander Mattingly. The astronauts in the spacecraft kept tabs on his health by occasionally asking, "Are the flowers in bloom yet in Houston?" The flowers never did bloom. Mattingly, a tall, partly bald man, took a seat next to Brand at the CAPCOM console; he had just run the entire checklist through the CMS/LMS Integrator and presumably knew it better than anyone else, so, in violation of the usual procedure, he was to read it up. Brand gave him the microphone.

The readup to Swigert took almost two hours. "O.K., let me take it from the top here," Mattingly began briskly. "It assumes that we're getting LM power to Main Bus B in the command module." He continued, "And the first item, then, after you get ready to start this checklist, is to install lithium-hydroxide canisters inside the command module.

. . . On Panel 8, we want to turn the floodlights to 'Fixed.' "

Swigert, who had already fallen behind, said, "O.K., wait a minute here. You're going too fast."

Mattingly's main concern throughout the reading was to see to it that Swigert, who was exhausted, got things right. Mattingly slowed down, reading the checklist line by line and pausing at the end of each line to wait for Swigert to read it back for confirmation. For accuracy, Swigert wanted to copy every word, using few abbreviations, and this made the reading go even slower. The word used most frequently over the next three hours was "O.K.," which prefaced each remark to confirm the last one.

"O.K. On Panel 250, circuit breaker, Battery A power entry, and post-landing closed," Mattingly voiced up.

"O.K. Panel 250, CB Bat A power entry, and post-landing closed," Swigert repeated.

"O.K. The same for circuit breaker, Battery B power entry, and post-landing," Mattingly went on.

"O.K.," Swigert repeated.

The material was detailed and tedious, and Swigert frequently had to ask Mattingly to repeat items. Once, Swigert broke off and said, "Ah, Ken, I—I didn't get it. We had to change omnis again." The spacecraft still rolled unpredictably, so that its omnidirectional antennas often lost contact and Swigert's voice trailed off. Mattingly and the flight controllers following the transcript had to be constantly on the watch for errors, and, even so, as time went by errors inevitably crept in. Once Swigert erroneously copied down that a heater for the guidance platform should draw power from Main Bus A, which wouldn't be powered up until later, and Mattingly had to correct him, telling him to plug it into Main Bus B.

After Mattingly read up the part of the checklist for jettisoning the LM, the pattern of the reading changed, for from this point the flight controllers had been able to borrow more from the original checklist for the mission, and accordingly Mattingly could tell Swigert, "Okay, on page nineteen, delete items one through five," or "Cut from page three to page six." This saved on penmanship for Swigert, who was getting writer's cramp. Almost two hours after Mattingly started reading the command-module procedures, he reached the end. "Thank you. This does it, huh?" Swigert asked, exhausted. He wanted to know whether Mattingly had had any trouble with the checklist in the simulator, and Mattingly said there had been no problems, though the pace was bound to be wearing between the jettisoning of the service module and that of the LM.

When Mattingly was through with Swigert, he had to start over again and read up procedures for the LM to Haise. These were shorter, and he finished within an hour. It was almost 11 P.M., just six and a half hours before the astronauts would start putting the checklist to use. For the next two hours, the ground made no calls to the astronauts, in the hope that they would get some rest. On the ground, most members of the White Team went home to get what sleep they could before taking over the Control Room for the final descent. Other members of the team continued to fret about details. In their view, the checklist was not really finished; that would require another three months.

At two in the morning, Lousma, who had replaced Mattingly and Brand on the CAPCOM console, called up to the spacecraft. He had some changes that the engineers wanted to make in the checklist. Swigert answered sleepily.

"How much sleep did you get, Jack?" Lousma asked.

"Oh, I guess maybe two or three hours. It was awful cold, and it wasn't very good sleep," Swigert answered.

"You plan to try to get any more?" Lousma asked.

"Well, if I get everything done, I'll try, but I'll tell you, it's almost impossible to sleep," Swigert said. "All of us have that same problem. It's just too cold to sleep. . . . We'll try to sleep, but it's just awful cold."

Lousma's call had awakened Haise and Lovell, too, and they were up now, rubbing their hands together for warmth. Lousma read up a couple of the checklist changes to Swigert, and then asked to talk to Haise; he had a couple of changes for him, too. Then Lousma had another change for Swigert, who had gone back to bed. Lovell, however, had had enough. He came on the radio saying, "O.K., Jack, this is Jim. I just want to make sure that any

. . . of the changes to the checklist that come up, you make sure that they're absolutely essential. When we don't have procedures we can only do it one time, and we can't make changes at the last minute. . . . Unless the changes are really essential, don't bother sending them up." At times, Lovell seemed more fretful than the others, perhaps because of his responsibilities as commander.

After Lovell's outburst, Donald K. Slayton, the Director of Flight Crew Operations, came on the line. Slayton was one of the few men on the ground who were allowed to interrupt the CAPCOM, and he did so now, evidently, because he thought a strong hand was needed. He said, "I know that none of you are sleeping worth a damn, because it's so cold, and you might want to dig out the medical kit there and pull out a couple of Dexedrine tablets apiece." This was a prescription the Flight Surgeons had been considering for some time but had been reluctant to suggest, because Dexedrine is a powerful stimulant, which leaves one severely let down when it wears off. The astronauts were reluctant to take the Dexedrine, too, but they dutifully told Slayton—who, they knew, had a paternal interest—that they would do so. However, they did not take the pills until a couple of hours before splashdown, when they were so exhausted that the Dexedrine had no appreciable effect.

When Slayton had returned the microphone to the CAPCOM, Lousma said, "If we could figure a way to get a hot cup of coffee up to you, it would taste pretty good about now, wouldn't it?"

"Yes, it sure would," Lovell said. "You don't realize how cold this thing gets."

Lousma replied, "Hang in there. It won't be long."

At two-thirty in the morning—a bit less than ten hours before splashdown—Lousma radioed, "O.K., Skipper, we figured out a way for you to keep warm. We decided to start powering up the LM now." The night before, the TELMUs had concluded that matters were going so well that they could loosen up some more on their consumables. The astronauts were told to turn on some of the equipment in the lunar module now, including the window heaters, which spread warmth rather in the manner of the defrosters in a car.

All three astronauts were up and about. "It's going to be an interesting day," Haise radioed. "The earth's a lot bigger, and the crescent is a lot more pronounced than it was yesterday."

The sun was shining brilliantly through the overhead window. Lovell remarked that it already seemed to be getting a little warmer in the LM. Lousma answered that duckblinds were always warmer when the birds were flying.

The White Team took over the third-floor Control Room shortly after four o'clock Friday morning, but most members of the team had come in long before that. Aaron had arrived a little after two to figure out what he would do if the command module used too much power (he called this kind of thinking "playing 'what if' games"), and Deiterich had come in at about the same time to draw up alternative maneuver PADS in case anything went wrong with the reëntry. When Reed arrived, around two-thirty, he went straight to the FIDO's console to look at the latest vectors. Because of the continued shallowing of the trajectory, the RETROS had definitely decided on one more midcourse correction—MCC 7—to be done about four hours before entry interface. The idea was to alter the trajectory by first tilting the spacecraft and then firing forward, and since the shallowing was even greater than Reed had expected, he and Deiterich now decided to increase the burn from one and a half feet per second to two feet per second; the effect of the burn would be to steepen the reëntry angle.

At four, Kranz officially took over. His earphones had the longest cord in the room, because he liked to pace back and forth at tight moments. He would be doing a great deal of pacing during the next eight hours. Still, he looked surer of himself and more determined than he had the night of the accident; the square set of his jaw might have been that of a commander who had been routed in a sneak attack and had now returned to the scene with what he believed to be an overwhelming force. He claimed later that he had had no worries. "In our preparations, we never believed we couldn't get the men back," he said. "I thought that as a group we were smart enough, and clever enough, to get out of the problem." However that may have been, almost every one of his flight controllers would have a problem during the next eight hours that he wasn't at all sure he could get out of. Most of the flight controllers were closer than Kranz to specific parts of the spacecraft, and were therefore apt to be more concerned when something went amiss. Kranz was generally able to take a longer view, and in this he may have been helped by his position in the Control Room—in the third row, a little behind and a little above the others.

Lovell said later that he could never forget for an instant that "we had a dead service module, we had a command module but it had no power in it, and we had a lunar module that was a wonderful vehicle but it didn't have a heat shield." The spacecraft was indeed in what Lovell called "an unusual configuration," and there wasn't much time to set it right. It was only fifty-eight thousand miles from earth now, and its velocity had increased to fifty-nine hundred miles an hour. In the next eight hours, this figure would more than quadruple. Contemplating the universe in man's usual egocentric way, Swigert told the CAPCOM that the earth—rather than the spacecraft—was whistling in like a high-speed freight train.

When Kranz took charge of the Control Room, he was surprised to find that the astronauts were aligning the LM's guidance platform on the sun and moon. According to the checklist, the LM's Primary Guidance and Navigation System, of which the platform was a part, was not supposed to be powered up—only the LM's secondary guidance system. What had happened was that during the night the TELMUs had found themselves to be what they called "fat on power," and had told the previous shift that it would be O.K. to bring up both of the LM's guidance systems. In one way, Kranz was pleased, because if the astronauts could get a good alignment on the LM's guidance platform now, they could transfer it to the command module later —as they had done in reverse the night of the accident—and this would save time during the most crowded part of the timeline. He didn't object to alterations in the checklist; it was not binding if a better way of doing something came along. What did bother Kranz was that, now that the primary system was up, the FIDO and the RETRO would want to use it for the MCC 7 burn, which was a couple of hours away. He suspected he would have an argument on his hands from the Control Officers, who would maintain that there wasn't enough fuel for that; they would want to stick with the more economical secondary system, as stipulated in the checklist. The issue was important. To save the command module's power and fuel, the LM would be performing all the spacecraft's maneuvers until it was jettisoned, six hours from then, and since there would be a lot of maneuvering, the Control Officers would be quite right in wanting to hang on to the LM's propellants.

Just as Kranz expected, he got a call from Reed, the Lead FIDO, who asked if there was any chance of using the primary system for the midcourse correction, because it would make for a more precise burn. Kranz said that he would consider it but that first he had to check with the CONTROL. To his surprise, the CONTROL on duty (the Lead CONTROL, Harold Loden, was not present) gave his approval. Kranz then checked with Russell, the Lead GUIDO, who also went along with the idea, so Kranz gave Reed permission to use the primary system, though he wasn't sure he had heard the end of the matter. Kranz liked to arrive at decisions by consensus, but as he asked a question here, answered one there, and passed an instruction to the CAPCOM to transmit to the astronauts, there could be no doubt that he was the center of the whole operation. As a rule, the conversation on the loop was terse, with long silences. Kranz's eye would rove around the room, and every once in a while he would seek information from one man or another. Once, he broke the silence to ask the CONTROL how he was coming; the CONTROL answered that the astronauts had already begun to use the LM's primary guidance system to hold the spacecraft at the proper attitude. After another silence, Kranz asked Deiterich if he was satisfied with the way the crew had stowed equipment in the command module for balance the day before; Deiterich said he was.

After another, longer silence, Kranz asked Peters, the Lead TELMU, how he was doing; Peters replied that he was doing fine, though the LM's batteries were beginning to shade off a bit as a result of the LM's early power-up.

So far, the LM had not begun to supply power to the command module, apart from recharging one of its reëntry batteries. A little after five in the morning, Swigert entered the command module to start drawing a little electricity from the LM—enough to turn on some of the heaters to warm electronic equipment inside the command module and, in particular, to de-ice the thrusters on the outside. (The GNCs, who were worried about the thrusters, had got further with the TELMUs than with Aaron.) The transfer of power to the command module began about an hour earlier than the checklist specified because, in addition to the LM's new found fatness, Peters, the Lead TELMU, had suddenly realized it wouldn't cost the LM any more of its water for cooling electronics—water being even more of a pacing item in the LM than electricity—because the command module would cool its own instruments with its glycol system. As the circuits were not set up for the LM to power the command module—Lovell's worries on that score had been quite reasonable—Aaron and Peters had thought up a way to trick them. Accordingly, Swigert plugged an entry battery into Main Bus B and drew just enough power so that the command module was in fact feeding a little to the LM; then, with the circuits between the two modules established, Swigert reversed things so that electricity from the LM was now flowing into the command module's Main Bus B. Swigert was relieved that this bus, the first to fail after the accident, proved to be in good condition.

For the first time since the accident, he felt as if he had something to *do.* The night of the accident, just after the command module had been powered down, Swigert told the other two that it was up to them to get the spacecraft home; now it would be largely up to him. As Swigert took his seat in the command module, Mattingly took a seat in the Control Room next to the CAPCOM. Although Mattingly had been working on the simulations to verify the checklist almost incessantly for the past three days, he looked spotless and crisp. At about the time that Swigert and Mattingly took their seats, Kerwin relieved Lousma as CAPCOM.

With all the early power-ups resulting from the LM's surplus, Kranz now thought it wise to warn the astronauts that under no circumstances were they to jump the gun and power up the command module on its entry batteries before the time stipulated in the checklist, two and one half hours before splashdown. The next step was to plug the heaters for the command module's thrusters into Main Bus B, to start de-icing them; and in the Control Room Kranz alerted Peters that at any minute he could expect to see the drain on the LM's batteries. This would be the first use of the new circuitry. Peters watched for it on the LM's telemetry, which was powered up now—but the drain did not show up, and some of the flight controllers worried that the reversal of the circuits had not worked after all. The trouble, though, was that Swigert hadn't pressed the switch yet because he was having difficulty deciphering the checklist he had copied the day before. "Either I copied the circuit breaker wrong, or—I can't read it . . . ," Swigert said. "Would you give me that again? I just can't read my own handwriting." (Even so, he said later that he was glad he had taken the trouble to write out the checklist longhand, without abbreviation, or he might have had even more trouble.) On the ground, Mattingly put his finger on the proper place on the checklist to help Kerwin, who read up the correct instruction. At length, Peters reported to Kranz that he could see the drain on the LM's batteries, and this was taken as evidence that the command module's thrusters were being warmed.

By six-fifteen, the glassed-in gallery at the back of the Control Room was beginning to fill with visitors who wanted to be present for the mid-course correction. This time, in addition to the NASA brass, there were a number of congressmen. The burn, less than an hour away, would be a small one, done by the lunar module's thrusters, which would fire forward to alter the trajectory. One reason for the burn was that Deiterich wanted to make sure the astronauts landed as near the recovery ship as possible; the sooner the astronauts were picked up the better, if only because Tropical Storm Helen was still charging erratically about the Pacific. The spacecraft might have come in all right without the burn, but on its present course it would come dangerously close to the shallow top of the corridor, and Deiterich wanted to center it against any further shallowing—if the rate increased, it might *not* stay within the corridor. Indeed, Reed was then finding that the trajectory was a good deal shallower than it had been when he arrived, four hours earlier, and he still couldn't figure out why. He and Deiterich therefore decided to increase the burn from two feet per second to three. Even with the increase, the burn was still small enough so that it could be done with the LM's thrusters. However, they would have to fire for twenty-three seconds—a long time for thrusters—and Loden, the Lead CONTROL, was worried about running out of fuel.

Loden, wearing a bright-yellow shirt, had just taken over the CONTROL console; he had been working elsewhere on some other matters and therefore had not been involved in the decision to use the LM's primary guidance system for the burn instead of the more economical secondary system. The astronauts had been using the primary to hold their attitude, and it was on an automatic setting, so the thrusters fired short bursts every time the attitude strayed. Since the LM had to juggle the command and service modules as well, the use of fuel in the automatic setting was more than Loden could tolerate. He complained to Kranz; he said they should switch over and hold their attitude manually as they waited for the burn. Kranz agreed. Then, when it appeared that some trouble in the primary guidance system could not be cleared up, Loden suggested that they do the burn with the secondary system. "Do the burn in secondary?" Kranz asked. The argument he had avoided earlier was about to start up. Reed, the FIDO, gave a distressed laugh and asked to have matters left the way they were. Kranz was quite good at settling arguments. He tended to listen quietly, ask a question or two, and then suddenly announce a decision. In this case, after a brief discussion, he told Reed no.

The CAPCOM passed the word to the astronauts. With the burn being done by the less accurate secondary guidance system, the Recovery Officers badgered Aaron to give them a commitment about how much power would be left in the entry batteries after splashdown; the spacecraft's radio beacon might have to beep a long time before it was found. Aaron, however, could give them no firm figure.

Although the MCC 7 burn was the smallest the astronauts had done, they had the most trouble with it, for exhaustion and dehydration were wearing them down. The flight controllers had to watch the astronauts' every move closely. On the telemetry screen at his console, Russell, the Lead GUIDO, was concentrating on following some data that Lovell was punching into the LM's computer for the correction. The numbers had been prepared by Deiterich, and Russell had a copy of them. Suddenly, Russell spotted on the telemetry screen the code number "P 40" where a "P 41" should have been. Lovell, who had stinted himself on water even more than the two other astronauts, had erroneously ordered the computer to use the LM's big DPS rocket instead of the thrusters. "Should be forty-one, I believe, Flight—not forty," Russell said quietly to Kranz, and Kranz told the CAPCOM to take note.

The MCC 7 burn caused trouble almost up to the moment the rockets were fired. In bringing the spacecraft to the attitude for the burn, Lovell rolled it sixteen degrees in the wrong direction. Deiterich, the Lead RETRO, didn't notice the error until he heard someone else mention it over the loop. Then, before he could say anything himself, he heard a second flight controller say that the error was unimportant, because the midcourse correction wasn't sensitive to being sixteen degrees out of attitude. "That griped me. During a landing, everybody in the Control Room thinks he's a RETRO," Deiterich said later. He came on the loop and said the burn certainly *was* sensitive to being sixteen degrees out of attitude. This was not the only time the astronauts made such an error. Moreover, the people on the ground seemed to catch their increased proclivity for errors. When Kerwin was telling Lovell how much time there was before the midcourse correction, he gave him a mark of ten minutes and then corrected himself, for the burn was only nine minutes off. Like Brand before the PC +2 burn and Lousma prior to the MCC 5 burn, Kerwin had been looking at the wrong electronic clock at the front of the room.

The burn itself, however, went off smoothly, and almost immediately Reed began checking vectors. He was able to get a hack on the trajectory sooner than before, because the spacecraft was approaching earth so much faster and the trajectory was beginning to curve slightly as it did so, making more of a difference between vectors. Besides, the spacecraft was near enough now for smaller radar dishes in the tracking network to get a line on it, and therefore the quality of radar data available to Houston was better. In a few minutes, Reed reported to Kranz that the spacecraft had apparently been tweaked back to the center of the corridor, despite the use of the less accurate guidance system. The spacecraft was only forty-one thousand miles away now, and its speed had increased to seven thousand miles an hour. The astronauts reported that the earth almost filled its window.

The real business of getting home was about to begin.

As Haise and Lovell, in the lunar module, maneuvered the spacecraft to the correct attitude for jettisoning the service module, Swigert, in the command module, made preparations for firing the explosives, called pyros, that would sever the service module from the rest of the spacecraft. Swigert was flicking a number of switches in front of him to open the circuits leading to the explosives—an operation called arming the pyros. It was risky, in the same way that taking off the safety catch of a gun is; if Swigert wasn't careful, the pyros could explode if he pulled one more switch. Normally, the ground checked the pyros before the arming, but at the moment it couldn't, because there was no telemetry from the command module. As a number of different pyros were on the same circuits, there was danger that the wrong thing might be jettisoned—the lunar module, for instance. In fact, the switch for jettisoning the LM was right next to the switch for jettisoning the service module, and the previous day Swigert, afraid he might press the wrong one, had put a piece of tape over the LM-jettison switch and attached to it a slip of paper bearing, in big red letters, the word "NO." Then he got Haise to verify what he had done. Now Haise, who wanted to be sure nothing went awry, and didn't want to be in the wrong module if it did, floated into the command module to help Swigert with the jettison. He asked Swigert whether he shouldn't check with the ground before arming the pyros, but Swigert, pointing out that there was no telemetry, told Haise just to stand by and put his fingers in his ears.

The jettison involved all three astronauts. Haise returned to the lunar module, and he and Lovell fired the LM's thrusters so that the spacecraft moved forward, with the service module at the front, at half a foot a second; then Swigert, in the command module, fired the pyros to separate the service module; and, finally, Lovell and Haise reversed the thrusters so that the service module and the rest of the spacecraft parted, at half a foot a second. Deiterich had made sure the speed was slow enough so there would be time to photograph the damage. The day before, Charles Duke and a photographer, in the simulator, had calculated that the best view of the service module would be from one of the windows in the command module, and, accordingly, Swigert was stationed there with the best camera, a Hasselblad equipped with a telephoto lens. But now when he looked out the window, his camera at the ready, he saw nothing but black sky. Since a command module had never before jettisoned a service module while it was itself attached to a lunar module, no one had been able to predict exactly what would happen. When Swigert fired the pyros, the craft had rocked unexpectedly at the explosion—the astronauts later described the movement as a "rippling"—and none of the modules wound up quite where they were expected to.

Swigert heard excited voices from the lunar module, where the others had just caught sight of the service module. He heard the whir of Lovell's movie camera and the steady click of Haise's still camera, which had been loaded with color film. Neither of those cameras had the resolution of Swigert's Hasselblad. The service module—a squat silvery cylinder with a huge rocket nozzle at one end—was spinning fast. Lovell, who had been the first to spot it, said, "O.K., I've got her.... And there's one whole side of that spacecraft missing.... Right by the high-gain antenna, the whole panel is blown out, almost from the base to the engine." That was the first anyone knew of the service module's having burst open. The gash ran down its length like a long slice in the hull of a ship. Lovell strained to get a look inside the damaged hull, but by the time it came into view it was a hundred feet away, and it was rolling so fast that peering in was difficult. Two of the three fuel cells—canisters like depth charges—glinted as the sun struck them, but on the shelf supporting them there was such a welter of broken metal and jagged bits of Mylar insulation that Lovell couldn't make out whether the oxygen tanks were there or not.

Haise caught sight of the rocket at the rear of the service module. "It looks like it got to the [rocket], too. . . . Just a dark-brown streak. It's really a mess," he said. As the service module kept rolling and pitching, he saw debris hanging out of the gaping gash in its side. It looked to him as if the explosion had gone through several stages. As Deiterich listened to the reports on the force of the explosion, one thought filled his mind. He knew that the ruptured oxygen tank had been quite near the ceramic heat shield at the bottom of the command module, and now that the tank failure was shown to have been so forceful, he was gripped by the fear that the command module's heat shield had been damaged, possibly even cracked. "I think everybody in the room had the same idea at the same time," he said later. "Everybody knew where the oxygen tank was. Nobody said a word about it. There was nothing anybody could do." Up in the spacecraft, the same thought was occurring to Haise.

Lovell shouted to Swigert that he should bring his telephoto camera into the lunar module; Swigert hurried in and practically climbed on Lovell's shoulders to get his shots. Because the service module was tumbling, the damaged part was usually obscured by shadows so dark that it was impossible to photograph the wreckage, and because the lunar module itself was rolling, the service module wasn't always in sight. Swigert had to run back and forth as it kept popping up at one window and then another. The astronauts wished they could steady their attitude for photography, but, in order to obtain the best tracking data, Deiterich had prohibited use of the thruster jets. For a time, they lost sight of the service module behind the command module, and when it reappeared it was over four hundred yards away and receding rapidly. The moon was in the background now, glaring white. They shot away with their cameras until the service module was the merest speck in the distance, the way shipwrecked sailors might keep gazing at the spot where their boat had sunk.

Kranz did not spend much time thinking about the explosion and the gash in the service module, for he regarded the mission as a flight of stairs to climb, and now that he was close to the top he didn't want to concern himself with anything that had happened on the bottom step. The time was approaching when Swigert would start drawing electricity from the reëntry batteries to power up the equipment that the command module needed for reëntry. Because the LM was supplying power to Main Bus B, the reëntry batteries would initially be plugged into Main Bus A, and Aaron now told Kranz he would like to get Swigert to test Main Bus A by plugging one of the reëntry batteries into it and taking a voltage reading. Kerwin passed the request to Lovell, who shouted it through the tunnel. (It would be some time before the command module's radio was powered up.) A couple of minutes later, Swigert passed word back that he was getting a reading of two amps on the bus, and Kerwin said that sounded good to him.

It didn't sound good to Aaron, for it meant that some bit of equipment in the command module was on when it should have been off, and until he found what it was it would continue to draw current that would be badly needed after splashdown —for the radio beacon that would signal the craft's location and for other, unforeseen circumstances. "We'd blown half our margin. That really threw us," Aaron said later.

Without telemetry from the command module, it was almost impossible to find out what piece of equipment had been mistakenly left on. It was precisely to avoid this type of situation that Kranz had earlier got Swigert to read off the position of each of the switches in the command module. Now there wasn't time to do that again. And at this point Aaron didn't particularly want Swigert to know he was worried about the drain, for the solution might yet prove to be a simple one and he would have alarmed the astronauts unnecessarily. Aaron did spell out his fears to Kranz, however, and Kranz didn't seem unduly alarmed—the way a skipper on a ship wouldn't be concerned with a problem in the engine room; at least, until it was shown to be insoluble. He had half a dozen other problems to think about. He knew that his flight controllers were naturally nervous about problems in their own areas, and he had enough confidence in them to assume that they would find the trouble and fix it.

Such sang-froid on the bridge didn't make the problem on the lower deck any easier. Confident though Kranz may have been about his Lead EECOM, Aaron and the other EECOMs were "sweating blood," as Aaron said later. They tried to guess some of the most obvious items that might be on when they should be off, such as certain lights in the command module. Swigert reported that they were off. Next, the EECOMs suggested that some of the heaters for the command module's thrusters might have been mistakenly plugged into Main Bus A, but that was not the case. Since it looked as if it would be some time before the EECOMs came up with the right answer, Aaron asked Kranz to have the reëntry battery disconnected and avoid wasting any more power now.

As the time for the power-up with the reëntry batteries drew closer, Kerwin, who was anxious to get on with it for the sake of Swigert's nerves, brought occasional gentle pressure to bear on Kranz. "Shall I give 'em a go?" Kerwin asked. Kranz was determined not to do so before the checklist said to, and he was watching one of the electronic clocks at the front of the Control Room. He told Kerwin to wait. Aaron, who was still fearful that the batteries would be weak, like an automobile battery on a frosty morning, reminded Kranz that the astronauts should not draw very much power from them at first. He told Kranz he had a list of equipment that could be turned off if the batteries showed signs of flagging.

"What do you say now, Flight?" Kerwin asked.

Kranz waited until the electronic numbers indicated that there were precisely two and a half hours until splashdown. "O.K., go ahead," Kranz said.

In the command module, Swigert switched on the reëntry batteries and then shut off the power coming from the LM. Lovell and Haise shut down some batteries in the LM that were no longer needed. Then Haise joined Swigert to help him turn on more of the command module's equipment.

Now, with the command module's reëntry batteries on the line, the power-up could really get under way. And it was high time, everyone felt, because its computer had to be loaded, its guidance platform aligned, the LM jettisoned, and a number of other things done before entry interface, the point, four hundred thousand feet above the earth, where the first thin traces of the atmosphere begin. It was from this point, just two hours away, that the spacecraft would start its final descent. Even with this much time, getting the guidance and navigation equipment, the radio, and all the rest of the systems up and on the line would take some time, particularly in view of Aaron's caution about not overloading the batteries at first. Nevertheless, both the astronauts and the flight controllers seemed happier now. There was even some laughter. Swigert, as he pressed the switches, felt like a captain who had been given a new ship after his first one had sunk. Aaron felt the same way, and so did Seymour Liebergot, the EECOM who had been on duty when the command module was shut down; he was sitting next to Aaron now. About the first piece of equipment that Swigert turned on was the command module's telemetry, and soon the electronic displays on the EECOM's console flickered on. To Aaron and Liebergot, this was the moment when the spacecraft really came alive again. "You saw it was cold, and you could see it warming up," Liebergot said later. "It was like having the machinery in a ship's engine room suddenly come to life."

When Russell, the Lead GUIDO, took a look at the first telemetry coming from the command-module computer, he saw signs of trouble; there was what he called a "flashing thirty-seven"—the number thirty-seven, which was the code for a serious malfunction, flashing from a field of other numbers to attract attention. It made him wonder whether the computer had been damaged by its prolonged exposure to cold. However, the trouble turned out to be merely that Swigert had failed to hold down long enough one of the buttons he had pressed to get the computer started.

The telemetry continued to be weak and unsteady, so the INCO, the flight controller responsible for the spacecraft's radio, asked Kranz to get the astronauts to switch to another omni antenna. The change didn't help, and large parts of the telemetry were lost. Even so, the EECOMs began scanning their screens for clues to what might be causing the power drain. At length, they found the trouble: the astronauts had inadvertently turned on a couple of switches, among them one for part of a backup control system. Aaron asked Kranz to have them turned off. Swigert flipped the switches, without fully realizing the importance of the request.

With his margin regained, Aaron settled down to follow the power-up. On his console he had a sheet of graph paper on which he had drawn a descending blue line representing the maximum amount of power the astronauts could use at any time during the next two and a half hours and still have enough left after splashdown. As Swigert turned on more equipment, Aaron put dots on the graph to indicate the amount of electricity left, and he connected the dots with a pencil line. As the new line lengthened, he was reminded of some of the discussions of the last few days. When Swigert turned on the fans for the command module's lithium-hydroxide system, Aaron noted that the air in the command module was already clean—scrubbed by the mailboxes in the LM, just as he had told a Flight Surgeon it would be. The argument had cost him two amp-hours. A little later, when Swigert turned on the glycol cooling system for the electronics—the subject of a battle Aaron had lost to a GNC—Aaron couldn't help noticing that the electronics were already so cold that they wouldn't need any additional cooling. Aaron did not suggest turning off any of this unneeded equipment, however, because the power had already been accounted for. Besides, Swigert was already moving on to the next major item on the checklist.

Swigert saw he was getting to the point where he would have to align the command module's guidance platform. It was the most critical alignment of all, because the guidance system would have to keep the spacecraft precisely within the reëntry corridor, which was only one and a half degrees wide. As Russell had realized earlier, the big problem would be getting a precise navigational fix, preferably on the stars. But whenever Swigert looked out the window, all he could see was a blizzard of what he called "little fluffy white objects"—presumably nuggets of ice that had formed from condensed steam from the LM's instrument-cooling system. Kerwin thought it sounded as though Swigert would have to forgo the star check and settle for a less accurate one based on the sun and moon. Russell had been holding on to the maneuver PAD for the sun-and-moon check, in the hope that it wouldn't be needed. Deiterich felt it was not reliable enough for the kind of accuracy required this close to the earth. Kerwin asked for the PAD now, and Russell handed it to him. Since there was a greater possibility of the astronauts' landing off target if the sun-and-moon check was used, the Recovery Officers once again plagued Aaron for a commitment on the amount of electricity that would be left after splashdown.

The astronauts could still *try* to get the more precise star check, though, and Kranz asked Kerwin to begin sending up Russell's coördinates for positioning the spacecraft, so that it could find the guide stars, any two of which Swigert would need in order to get the fine alignment. In the meantime, Russell was sending information to update the spacecraft's computer, and after he sent the first of four uplinks he got no acknowledgment back to indicate that it had been safely loaded. Again Russell thought of cold damage. However, the main problem just then had to do with radio communication, which still had not been firmly established with the command module. Russell listened in on the loop over which radio technicians at NASA's network of tracking stations around the world could talk to each other. Some technicians at the station in Honeysuckle Creek, Australia, were having trouble with what they called lockup—locking their antenna onto the spacecraft's signal. They kept picking up the signal and then losing it. A technician with an Australian accent was repeating over and over again the refrain "We've got lockup. . . . No, we don't!"

Kranz asked the INCO if he had any idea what the trouble was. The best idea the INCO could come up with was that perhaps the lunar module was causing interference by getting between the command module's antenna and the station in Australia—a possibility that had never been anticipated, because lunar modules were not normally brought back to earth. The INCO suggested that the astronauts bring the spacecraft to a different attitude to get the LM out of the way. Russell was against this proposal, because the spacecraft was now at the proper attitude for the star check and he didn't want to lose it. Fortunately, the radio problem cleared up a little later, when the NASA tracking station at Guam took over communications with the spacecraft. (The INCO's guess about the LM's causing interference was afterward proved correct.) Now voice communication was established with the command module, and Kerwin could talk directly to Swigert without having to relay messages through the LM.

When Kerwin finished sending up the coördinates for the star check, Swigert was as skeptical as Kerwin and Russell had been about his chances of getting even one of the stars lined up in the command module's sextant. He peered through the sextant, which was set through the spacecraft's hull just above his seat and (unlike a similar instrument in the LM) could be swivelled without the need for moving the entire spacecraft. Swigert reported that the snowstorm of fluffy white balls glinting in the sunlight was still obliterating all the stars. He was already more than five minutes behind the timeline, and the earth outside the spacecraft window was getting bigger.

Swigert had already transferred to the command module the alignment from the lunar module's guidance platform which Lovell and Haise had set up earlier that morning. It was not a very reliable one, not only because it was already several hours old but also because errors could have crept in during the transfer. Lovell and Haise had shouted the angles through the tunnel, and if the spacecraft had rolled, pitched, or yawed before

Swigert could punch them into the command-module computer there would have been errors. Still, it would be easier for Swigert to tweak up an existing platform, crude as it may have been, than to start from scratch.

Swigert radioed down that wherever he turned the sextant the blizzard of ice particles still covered the stars. One of the flight controllers suggested to Kranz that they recommend a couple of guide stars on the side of the spacecraft away from the sun, where the spacecraft's shadow could cut the glare. He handed Kerwin the sextant angles for two such stars, Altair and Vega. Swigert swivelled the sextant toward Altair. Doing so was a departure from the usual way of aligning the platform, which was done by asking the computer to aim the sextant at a particular star, but Swigert felt that his present alignment was so crude that he didn't want to waste the time it would take the computer to guide the sextant to a position that could only be the wrong one. When he put his eye to the lens, he still couldn't see anything at all. Kerwin advised him to turn off a light, in the hope that he could see better. Meanwhile, Lovell, in the LM, who could see from the window that the earth was getting bigger at an alarming rate, shouted to Swigert to hurry up.

With the light off, Swigert found Altair easily. Occasionally, a fluffy speck of ice—black now, in the absence of sunlight—eclipsed the star, but otherwise Altair stayed firmly in view, a solid benchmark in the sky. As soon as he had it centered on the sextant's cross hairs, he ordered the computer to find the same star and then sat back to wait until the amount by which the computer was off—the platform error—appeared in the gauge. This had five windows for numbers, but when the five numbers appeared they were all zeros; Swigert said he got "five balls." There was no error at all. Haise's and Lovell's rough alignment on the sun and moon and its transfer to the command module after several hours had been entirely accurate—an extremely unlikely circumstance. Quickly, Swigert found Vega and repeated the check; he got five balls there, too. Suddenly, the astronauts were five minutes ahead of the timeline.

For the first time, Kerwin said later, he let himself think that they were going to make it.

As soon as the command module's guidance platform was aligned, Lovell, alone now in the lunar module, began to maneuver the spacecraft to the right attitude for the jettison of the LM. Following the method that Charles Duke had found to be practicable in the simulator, Lovell rolled, pitched, and yawed the spacecraft so that he could see the succession of guide stars out the window, as if they were buoys leading him to port; since he wasn't looking at them through a sextant, he wasn't as bothered by glare as Swigert had been. As it turned out, the guide stars were less like buoys leading to a safe harbor than like buoys skirting a shoal, for the path he was following from star to star brought the LM's guidance platform close to gimbal lock—in spite of the fact that the series of stars had been changed since Duke's trials in the simulator. In the Trench Deiterich, Reed, and Russell were puzzled. Whenever Lovell was about to jam the gimbals and lose the platform alignment, Loden, the LM's Control Officer, called out a warning to Kranz, and Kerwin relayed it. Then Lovell would back up a bit to pick a new route. It was slow going, and Lovell was getting impatient. Entry interface was just an hour and a half off; they were only about eighteen thousand miles away now, and their speed had increased to almost eleven thousand miles an hour.

Though no one was aware of it, the command module's guidance platform—more important now than the LM's—was being maneuvered close to gimbal lock, too. The flight controllers, with their attention focussed on Lovell and the LM's platform, forgot all about the alignment that had just been set up in the command module, until Swigert called out that it was in danger of being lost. So much time and effort had been invested in aligning this platform that some of the flight controllers thought Swigert must be joking, but he wasn't. He shouted to Lovell to stop the maneuvering, and then he called some directions through the tunnel, rather as if he were a pilot guiding a big ship around a sandbar. "A little more pitch! A little more pitch!" he called, and when the danger was past he shouted, "Now you can begin to roll!" Lovell grumbled to Kerwin that he had picked a lousy attitude for jettisoning the LM, and Kerwin told him to take his time, since they were a little ahead of the timeline.

Lovell, however, was in no mood to wait. The earth was looking less and less like an ethereal celestial body and more and more like a big solid landmass dead ahead. When at last he got the spacecraft to the proper attitude, he put the LM's Primary Guidance and Navigation System on its automatic setting, so that the LM would be maintaining the spacecraft's attitude until it was jettisoned. Then he told Kerwin he was planning to bail out of the LM. Kerwin said he couldn't think of a better idea. Before scrambling through the hatch, Lovell took a last look at the LM's cockpit, crammed to the ceiling with debris collected during the six days of the flight. He thought that it looked like a packed garbage can. He shut the lid, the LM's hatch, behind him.

The three astronauts were together now in the command module. Swigert switched on the oxygen surge tank that Kranz and Liebergot had hastily ordered isolated the night of the accident, and the command module's own oxygen flooded the cabin. On the ground, too, everyone's attention turned to the command module. Kranz asked Aaron how it was doing on power, and Aaron, after consulting his chart, said that it was doing well—that if things kept on going the way they were going then, there would be plenty left after splashdown. Moments later, however, he had to amend his estimate, for he found another unexplained drain on the batteries. After some digging around in the telemetry, he discovered that Swigert had left on the power that ran the sextant.

With this remedied, the margin was in good shape.

Buck Willoughby, the Lead GNC, had a more serious problem. Willoughby's telemetry was indicating that at least two of the command module's thruster jets were still cold, though Swigert had turned on their heaters a couple of hours before. As the GNCs had told Aaron a day and a half before, if fuel had become frozen inside their nozzles, they mightn't be able to keep the command module at the right attitude during reëntry. Willoughby told Kranz he would like to test-fire the thrusters now. Kranz, however, didn't want to risk knocking the spacecraft out of the jettison attitude that Lovell had had such a hard time setting up. Kranz asked Willoughby if he couldn't wait to test the thrusters until after the LM was separated, a wait of less than half an hour.

Kranz needn't have been so meticulous about the spacecraft's attitude. Deiterich, who had been uneasy since the unexpected brushes with gimbal lock, now checked his telemetry to make sure the spacecraft was at the right attitude for jettisoning the LM, and he thought that things didn't seem quite right. Under the best of circumstances, nothing upsets a RETRO or a FIDO more than an attitude error, but this time there was the added problem of aiming the cask of radioactive fuel that Deiterich had promised the Atomic Energy Commission he would set down off New Zealand. Deiterich asked Reed, the Lead FIDO, what the correct attitude angles were supposed to be, and together the two flight controllers discovered that Lovell had brought the spacecraft's angle of yaw ninety degrees from where it should have been. In addition to explaining the close calls with gimbal lock, the error meant that the lunar module was going to be jettisoned toward the northeast instead of the southeast.

Reed took the matter up with Kranz.

"How far off are we?" Kranz asked.

Reed told him, and added that the error should be corrected.

Kranz replied that any correction would be difficult now, because the lunar module had been closed up and it would take too much time to open it again. Kranz asked Reed exactly how the attitude error changed the situation, and Reed told him it would alter the point where the spacecraft hit the atmosphere.

Kranz thought that Reed was talking about the point where the LM would hit the atmosphere, and asked him if the radioactive-fuel cask was what he was worried about.

Reed said no—that he was concerned about where the command module would hit the atmosphere, for if the LM went off course to the north, the command module would go off course to the south. It would, in fact, go farther off course than the LM, for it was lighter and so would get a harder push when the two modules were blown apart by the air in the tunnel. Reed said that this error could affect the astronauts' passage through the corridor and where they landed, but Kranz disagreed. He told

Reed he felt that the error in the command module's trajectory could be corrected later, so he was inclined to leave things alone now. However, he wanted to hear what Deiterich had to say.

Deiterich was still worrying about the fuel cask, possibly because he felt the gaze of the A.E.C. man on the back of his neck. He rapidly figured out where the cask would come down under the new circumstances, and he was able to assure the A.E.C. representative that it would still come down in deep water. If there were any chance of getting into the LM easily, it would still be better to make the correction now, but, like Kranz, he was worried about the time it would take to reopen the LM and bring the spacecraft to the correct attitude. He knew the chances of doing this were slim, so he said to Kranz, "Lookit, if he's closed out in the LM, let's let him alone. He'll be behind."

Kranz said that Lovell *was* closed out in the LM.

"O.K., we'll just *buy* it," Deiterich said. He was not particularly pleased with what he was buying. He had already had enough problems with the trajectory, and it wasn't out of the question that he would have some more.

Lovell, who knew nothing about his attitude error, was working on another problem: checking the integrity of the command module's hatch—a matter that had come up the day before, when Lovell rejected the idea that he and the others wear their space suits during reëntry. To make a complete check, the hatch had to be shut and air had to be vented from the tunnel, which would be closed off at both ends. If there was a leak, Aaron would detect it through an increase in the oxygen flowing from the surge tank to compensate for the outflow. The check would be complicated by Deiterich's plan to blow the two craft apart with the air in the tunnel; he wouldn't be able to do this if Swigert vented much air. The RETRO and the EECOM had discussed the problem, and Deiterich had agreed to let Swigert vent half the air; that would still leave enough pressure for a good jettison.

After Swigert opened the valve to let the air out of the tunnel, it took ten minutes for the tunnel pressure to drop the right amount. Aaron waited a couple of minutes longer before checking, so that any leak would have time to show its effects. Then, however, he found that the surge tank was pumping more oxygen into the cabin than it ought to be. Was the hatch leaking, then? He made no mention of the possibility at that moment, for he didn't want to put everyone in what he called a "worry mode" until he had checked out a couple of other explanations. Swigert noticed the extra oxygen flow himself, though.

Within a minute, Aaron found out what the trouble was. Until recently, the lunar module had been providing the command module's oxygen, and the LM's oxygen was set to a slightly lower pressure than was usual in the command module. Consequently, it would take an extra surge of oxygen to bring the command module's pressure up to normal. It was this extra surge that Aaron had mistaken for evidence of a leak. Three minutes later, the excess oxygen flow stopped, and Aaron relaxed.

Almost as soon as the oxygen flow had steadied, Willoughby, the Lead GNC, came on the loop to tell Kranz of a new difficulty: the command module's guidance platform was once again in danger of gimbal lock. There was almost nothing the astronauts could do about it this time, since the spacecraft's attitude was being controlled automatically by the LM, which was inaccessible. Shortly, Loden, the CONTROL, warned Kranz that the spacecraft's wobble was going beyond the limits of what the LM's guidance system could handle.

Kranz decided that the astronauts had better get rid of the LM as soon as possible. First, however, he had to make sure of a few things. He asked Aaron how the command module's cabin pressure was doing. Aaron said it was fine. The LM was not needed to plug a leaky hatch. Next, Kranz turned to Willoughby, who had been afraid earlier that all the command module's thrusters might not fire, and asked him whether he still wanted to test-fire them before the LM was jettisoned. Willoughby replied that the thrusters looked warmer now, so he didn't think there would be a problem. The LM would not be needed to spin the command module into the corridor. Then Kranz turned to Loden and asked if there were any other problems with the LM that he should know about. There weren't. "O.K., let's punch off the LM," Kranz said. But before he could give the order, he had to check with each flight controller in turn—an exercise he called "going around the horn," which he was required to do before ordering any irreversible action. He ticked off the names rapidly—RETRO, FIDO, GUIDO, CONTROL, TELMU, GNC, EECOM, Surgeon, and INCO—and the answers came back just as fast: "Go," "Go," "Go."

The entire roll call took about ten seconds, and when it was finished Kranz told the CAPCOM to tell Lovell he could separate the LM as soon as he was ready. In spite of all the "Go"s, there was still a problem; Apollo 13 would be iffy until the end. When the umbilicals that had been supplying power from the LM to the command module were disconnected, a short circuit might cut the LM's power supply, turn off its guidance system, and allow it to collide with the command module or go far off course with its radioactive cargo. But discussion among the EECOMS, the TELMUS, and the Grumman engineers had convinced the controllers that the risk would have to be taken, and the order was given for the pyros to be fired. It was agreed that when the jettison was done, the LM's guidance system should insure that the two craft would not come together again, and that the fuel cask would land in deep water.

At length, Lovell reported that the LM was gone. The LM Systems Officers felt a momentary pang; it would have been nice, they thought, if the LM could be brought back to earth. It had proved a better lifeboat than anyone had had a right to expect. Much later, a TELMU drew a cartoon—which he presented to the CM systems engineers—showing a jaunty lunar module prancing through the sky, dragging on a stretcher a sick-looking command module swathed in bandages, its tongue hanging out, a transfusion bottle dangling from its antenna. (Grumman also submitted to North American Rockwell towing charges for the approximately three hundred thousand mile trip.) Now, as the actual LM spun off into the distance, its golden legs kicking in space and the sunlight glinting on its silvery hatch, Haise, its pilot, snapped photographs until it became an iridescent speck in the distance. The separation had rocked the command module slightly, and Kerwin radioed up that the astronauts had better watch out for gimbal lock. Kranz's attention returned to the command module's hatch, now that there was no longer a lunar module to protect it. "How's your pressure doing, EECOM?" he asked. Aaron said it still looked all right.

The command module was flying alone now, the way it would have been if no accident had occurred. The astronauts were back on the original timeline for the mission, though they were four days ahead of their original schedule, and though heavy equipment from the LM was stowed where moon rocks should have been. Entry interface was an hour away. The astronauts, travelling at about fifteen thousand miles an hour, were only about eleven thousand miles from their target on earth, and considerably less than that from the earth itself, which was big enough now so that it looked like a sea of blue outside a ship's porthole. Clouds scudded across it like waves. In the next half hour, the spacecraft's trajectory would curve around it like a cable winding onto a winch, and at a point immediately opposite the moon the spacecraft would drop like an anchor through the air and into the ocean, splashing down near the aircraft carrier *Iwo Jima.* That is, the spacecraft would do this if all went well.

Deiterich was sitting at his console preparing the final reëntry PAD. The rest of the flight controllers had all had their moments of worry since the preparations for reëntry began: Russell, the GUIDO, over the command module's alignment; Reed, the FIDO, over the attitude for separating the LM; Willoughby, the GNC, over the heating of the thrusters; and especially Aaron, the EECOM, over both the hatch's integrity and the extra two-amp drain. So far, Deiterich had seemed almost as cool and impassive as Kranz.

In order for the spacecraft to get through the narrow corridor that was its only safe passage to earth, it had to hit the atmosphere at an angle no steeper than 7.3° to the horizon and no shallower than 5.5°. Deiterich had assumed, for the reëntry PAD he was working up, that the spacecraft would hit the atmosphere at an angle of 6.51°, which had been the forecast the last time he checked the spacecraft's trajectory. That was steep enough so that Deiterich had stipulated in the reëntry PAD that the spacecraft should reënter the atmosphere at an attitude called "lift vector up;" that is, the command module's flat heat shield would tilt slightly upward, to make the spacecraft plane a little. Now Reed handed him the latest vector information, and Deiterich saw that the angle predicted for reëntry had shallowed out to 6.2°. It was the same unpredictable shallowing they had been seeing all along, and the new figure threw all Deiterich's calculations off. "I don't believe you, FIDO!" he said to Reed. Reed said later that Deiterich had almost scared him to death—just as Reed had frightened him.

Deiterich was in a very tight spot. In shallowing, the trajectory was approaching the angle 6.08°, at which the spacecraft, instead of coming in lift vector up, would have to come in lift vector down, to make the spacecraft dig into the atmosphere and so counteract the shallowing. Coming in with the wrong lift vector would affect the landing site—if, indeed, it didn't knock the spacecraft out of the corridor altogether—but Deiterich had no way of knowing whether the shallowing would continue at a rate fast enough to require a change in the lift vector from up to down. It had only a little over a tenth of a degree to go, and so far that morning it had already shallowed three-tenths of a degree. But there was no guarantee that the trajectory would continue to shallow at the same rate. "That's the kind of thing that really splits a RETRO," Deiterich said later.

Kranz kept asking Deiterich how he was coming with the reëntry PAD, since it was high time the contents of the PAD were transmitted to the spacecraft. Kranz, being one step removed, was exhibiting an almost unnerving cool, but it was his job to mesh all the last-minute details for reëntry, and the reëntry PAD was the only missing one. Deiterich, for his part, wanted to hold off on the PAD until the last possible minute; among other reasons, the PAD contained the exact times for automatically deploying the parachutes, and these times would be different depending upon which lift vector was used. Having the parachutes open at the wrong altitude could be as disastrous as any of the dangers that had been avoided thus far.

Up in the spacecraft, Swigert, too, was asking for the PAD, but Kerwin managed to put him off, saying that it would be a few minutes before they got it right up to speed. Deiterich, who felt he was beginning to run out of time, passed to Russell, the GUIDO, some of the reëntry information that would have to be up-linked to the spacecraft computer; however, he asked Russell to hold off sending it, because some of the data might have to be changed. Russell punched the information into his keyboard, so that the up-link was set up on his electronic screen, all ready.

Then Deiterich checked with Reed to see how the trajectory was doing. It did not seem to be shallowing as fast as it had been before, so Deiterich decided to gamble on a reëntry with the lift vector up, and he advised Kranz and Russell to transmit the reëntry PAD and the up-link the way they were.

Deiterich made the right choice. It was later

learned that the cause of the shallowing had been the constant boiling off of water into space from the LM's cooling system, and now that the LM was gone the shallowing had slowed down. The more the LM had been powered up, the greater had been the venting, and that was why the shallowing of the trajectory was unusually strong that morning. No one knew this at the time, even though everyone had been constantly aware of the continuous drain on the LM's water and the blizzard of ice crystals that the water made outside the window. The venting was on such a small scale that it had never caused trouble during the short hops of previous LMs down to the moon and back, so no one had realized that the negligible thrust it imparted to the spacecraft would build up appreciably during the LM's long flight back to earth. (Earlier, there may have been some venting from the service module's hydrogen tanks as well.)

Deiterich, who did not know that his troubles had flown off with the LM, determined privately to go ahead with the reëntry PAD and the up-links, and that he would not make the final decision about whether to change the lift vector until just before the time of the moonset check, which the astronauts would carry out two minutes before reëntry. In the meantime, he began figuring out the changes that would have to be made in the timing for the parachutes if the shallowing increased at the last minute.

The spacecraft was swinging behind the earth now, like a ship turning to enter port. The earth had long since ceased to be a ball in space, or even a sea of blue, and was now "land," in the sense that sailors returning to shore after a long voyage would use the word. Now that the final preparations had been made, the flight controllers had less to do and more time to think about what lay just ahead, so their conversation became increasingly terse. Kranz decided to take an informal trip around the horn. The EECOM, the GNC, the FIDO, and the GUIDO all said that things were "looking good." When Kranz called on the RETRO, there was no answer; Deiterich was totally absorbed in the trajectory.

Up in the spacecraft, the planet on which the astronauts were about to land had virtually vanished, because they were well around on its dark side now, away from the sun. Though the landing would take place close to noon Houston time, it was early morning in the mid-Pacific. In the dark, it was almost impossible to discern the earth's horizon, and this meant that the astronauts couldn't make the usual last-minute check of their trajectory by clocking the instant a given point on the horizon cut across the reticle etched in the spacecraft window. Instead, they would have to make a moonset check—the one Deiterich had thought up a couple of days earlier—and it was this that he was relying on for his final decision about the lift vector. The moonset check was a variation on the horizon check, and Deiterich liked it even better, because it "synched in" the movements of the earth and moon with the positions of the stars, and so fixed the trajectory even more precisely. Earlier, Swigert had brought the spacecraft to the best attitude for watching the moon, and now he had the moon lined up on the reticle in the window. As the spacecraft swung farther and farther around the earth, the moon gradually sank lower in the sky, until at last it set, revealing for an instant where the earth's horizon was. Just at the predicted time, the horizon bit a black arc out of the hard-white moon. It was not the fuzzy, diffuse sort of moonset that is seen through the earth's atmosphere. Swigert said later that the moon sort of blinked out. For the first time in six days, the astronauts were out of sight of the moon.

They now had less than a quarter of the earth to round. They were traveling at twenty-one thousand miles an hour, and they would be going twenty-four thousand miles an hour in ten minutes' time, when they reached entry interface, four hundred thousand feet above the ground. There the friction from the atmosphere would begin to heat up the spacecraft's heat shield. Nineteen seconds later, the resulting flames would make radio contact impossible. The heat would reach five thousand degrees, and for at least three minutes—until the spacecraft's speed had slowed and the temperature had dropped enough for communications to be reëstablished—no one would know how the astronauts were faring. As the time for interface approached, Deiterich and the other flight controllers, in view of what they had recently learned about the violence of the tank failure in the service module, were getting increasingly apprehensive about the condition of the heat shield. Up in the spacecraft, Haise worried, too—he regretted that he couldn't go out and take a look at it. Things had held together thus far; he hoped they would hold together a little longer. The flight controllers and the astronauts talked about inconsequential matters, among them the splashdown party that the flight controllers always have after a flight. Swigert told Kerwin he wished he could be there for it, and Kerwin said it would be a wild one. They talked a bit about the lunar module, too, for the TELMU had just reported that the LM's telemetry, which he had been monitoring, had suddenly stopped. Presumably, the LM had burned up in the atmosphere—the fate of any craft without a heat shield.

"Where did she go?" one of the three astronauts asked.

"Oh, I don't know. She's up there somewhere," said Kerwin. The astronauts and the flight controllers were all a little sentimental about the LM.

Just then, the TELMU reported another burst of telemetry from the LM, indicating that it hadn't burned up yet. "How about *that?*" said Kranz, with the only trace of astonishment he had betrayed since the accident. Like mariners, flight controllers prefer to think of favored craft as sailing on and on forever. After this short burst of telemetry, though, nothing more was heard from the LM.

Astronauts and flight controllers are generally

uneasy when it comes to expressing their feelings. Now, as the spacecraft raced on toward entry interface, there were long silences, and Swigert ended one of them by saying, "I know all of us here want to thank all you guys down there for the very fine job you did."

Lovell added, "That's affirm."

Kerwin replied, "Tell you—we all had a good time doing it."

It might have been the end of any agreeable guided tour.

The flight controllers picked at details like mother hens until the last minute. Aaron sent word to the astronauts that one of the reëntry batteries would give out at about the time the main parachutes opened but that they shouldn't worry, because there would be enough electricity in the other two. A little later, Aaron sent another message: If they landed off target and needed more electricity to power the radio beacon, there was a little extra supply stashed away in the small battery that powered the pyros. Kranz felt it was time to go around the room once more to see if everyone was ready for entry interface—not that there was much to be done now if anyone wasn't.

"O.K., fellows, last time around the horn," he said. "How we looking, GNC?"

"Good."

"EECOM?"

"Good."

"GUIDO?"

"Go."

"FIDO?"

"Go."

"RETRO?"

"Go, Flight," Deiterich said. The trajectory had continued to shallow until it was just a twentieth of a degree above the point where the lift vector would have to be changed; then the shallowing ceased.

"Good," said Kranz. He had made it round the horn even faster than before, for, with the LM gone, he hadn't needed to poll the TELMU or the CONTROL. Kranz said to Kerwin, "We're go, CAPCOM."

"We just had one last time around the room, and everybody says you're looking great," Kerwin said to Lovell.

Lovell said, "Thank you."

A little earlier, Swigert had told Kerwin he had a good bedside manner, and this may have been closer to the way he felt now, given everyone's unstated concern about the heat shield. In the first thin traces of the atmosphere, a purple-pink glow surrounded the spacecraft, moving backward from the shield in the manner of a gas fire being turned up slowly beneath a pot. Swigert said his mood fitted the colors. Shortly, the flames enveloped the capsule, so that the flight controllers lost contact with the astronauts. The flames trailed hundreds of miles behind the spacecraft, like the wake of a ship. The flight controllers stared silently at the front of the Control Room, where projected on the big television screen was a view—relayed by satellite from the *Iwo Jima*—of the sky the astronauts would be coming down through. The weather was nearly perfect. Deiterich gloated, for the storm that the weathermen had been warning him about for three days had moved off; it had moved, in fact, to the precise spot to which they had wanted him to shift the landing site.

After three minutes of blackout, Kranz put through a call to Deiterich to find out how much longer they had to wait. Deiterich said it should be over in another thirty seconds. At the end of the thirty seconds, there was still no word from the astronauts, and Deiterich began to get concerned. Thirty seconds later, the astronauts still hadn't reported in, and Deiterich was alarmed. Blackouts didn't always end on time, but this one was already excessively long. Kranz asked the Network Officer whether the spacecraft's radio beacon had been acquired. It hadn't been. Another thirty seconds went by, and Kranz asked Kerwin to put in a call to the spacecraft. "Odyssey, Houston. Standing by," Kerwin said. There was no answer. Everyone was beginning to despair.

Then, five seconds later, Swigert called in—"O.K., Joe." He sounded exhilarated and relieved. He was a minute and forty-five seconds late—one of the longest delays in any Apollo blackout. Kerwin told Swigert he read him. None of the flight controllers said a word.

It was nine minutes until splashdown. At twenty-four thousand feet, the two small drogue parachutes popped out to slow the craft, like sea anchors dragging against a current. On television, the flight controllers could see the spacecraft now. The drogues were reefed, but, as they watched, the reefs fell off and the drogues billowed. So good was the weather, and so close had the spacecraft come to its target, that this was the first time the flight controllers in Houston had seen this happen. Soon the drogues were released and three more small parachutes took their place, each one pulling behind it a vast parachute—the three main ones. These, too, were reefed, and as the reefs slipped out in two stages, the chutes appeared to grow like rocket bursts welcoming a ship to port. The astronauts touched down at 12:07 P.M. Houston time. The flight controllers cheered.

After the astronauts were safely aboard the *Iwo Jima,* the flight controllers applauded again and lit cigars, as they always do at the end of a flight. One of them mounted a stepladder and attached the Apollo 13 emblem to the wall alongside the insignia of other missions successfully brought home—it was a troika of horses dashing over a gray-and-white moon, with the sun just behind them and the earth, all whites and browns and blues, hanging invitingly in the background. Most of the flight controllers went on to the splashdown party, at the Officers' Club at Ellington Air Force Base, as soon as they had shut down their consoles. Kranz felt reluctant to leave the men he had been working with; it always took a little time to get away from a spaceflight, and the more demanding the flight had been the more time it took. Russell said later he was surprised at how long it took for his adrenalin to go down; the problems had been so vast in comparison with what he thought of as "the little gremlin-type problems" he normally got during a mission. Reed, too, expected to unwind at the party, but he did not. He was so cranked up that his mind wouldn't stop working. Five days later, he was still walking around in what he called a "shocked mode."

In spite of all those extra amounts of adrenalin, the flight controllers' party was not the "wild one" that Kerwin and Swigert had spoken of; in fact, Swigert didn't miss a thing. Everyone was almost too exhausted to move. Aaron had two drinks and was home in bed by six that evening. Deiterich drank a couple of beers, picked up his wife and took her out for a hamburger, and went home. Spencer, another RETRO, had a drink, went out for a pizza, and went home. Liebergot, the EECOM who had been on duty at the time of the accident, left the party early, too. Two weeks later, he said there had not been a single night when he hadn't dreamed about the undervoltage on Main Bus B.

When the astronauts boarded the *Iwo Jima,* a band struck up "Aquarius." After a short welcoming ceremony, Haise told the ship's captain that it was nice to be warm again, and Swigert said, "It was so damn cold." This was uncharacteristic talk for an astronaut, and a few weeks later, after he had had time to think things over in the warmth of the South Pacific and then Houston, Swigert was to deny ever having been particularly uncomfortable in the spacecraft. Haise would later liken the discomfort to that on a camping trip through the north woods without adequate clothing. Yet the frogmen who met them commented that they could still feel the chill of space through the command module's open hatch.

Belowdecks, in the sick bay, Lovell, Swigert, and Haise were given all the fruit juices they could drink. They didn't have the appearance of most of the returning crews the attending physicians had seen before, and a quick post-flight examination revealed them to be in worse shape than any other returning astronauts. They were unsteady on their feet and their reflexes were slow. They did worse than any of their predecessors on their orthostatic tests, which measured the pooling of blood at the feet—an aftereffect of space travel which is exacerbated by exhaustion. Their supply of electrolytes was significantly down, and so was their circulating level of adrenal steroids—secretions that provided an index of the extent to which the astronauts had been working from nervous energy. Before going to sleep, one of them told one of the doctors he just didn't know how much longer they could have kept on going.

Library of Congress Cataloging-in-Publication Data

Cooper, Henry S. F.
[Thirteen, the flight that failed]
Thirteen, the Apollo flight that failed / by Henry S.F. Cooper.
p. cm.
Originally published: Thirteen, the flight that failed. New York : Dial Press, c1972.
Includes index.
ISBN 0-8018-5097-5
1. Project Apollo (U.S.) 2. Apollo 13 (Spacecraft)—Accidents. 3. Space vehicle accidents—United States. 4. Space rescue operations. I. Title.
TL789.8.U6A5313 1995
629.45'4—dc20 94-39726
CIP